SpringerBriefs in Water Science and Technology

For further volumes:
http://www.springer.com/series/11214

Fei Liu · Guoxin Huang
Howard Fallowfield · Huade Guan
Lingling Zhu · Hongyan Hu

Study on Heterotrophic-Autotrophic Denitrification Permeable Reactive Barriers (HAD PRBs) for In Situ Groundwater Remediation

 Springer

Fei Liu
China University of Geosciences
Beijing
People's Republic of China

Lingling Zhu
Geological Publishing House
Beijing
People's Republic of China

Guoxin Huang
Beijing Academy of Food Sciences
Beijing
People's Republic of China

Hongyan Hu
Hydrogeology and Engineering Geology
    Prospecting
Institute of Heilongjiang Province
Harbin
People's Republic of China

Howard Fallowfield
Huade Guan
Flinders University
Adelaide, SA
Australia

ISSN 2194-7244   ISSN 2194-7252   (electronic)
ISBN 978-3-642-38153-9   ISBN 978-3-642-38154-6   (eBook)
DOI 10.1007/978-3-642-38154-6
Springer Heidelberg New York Dordrecht London

Library of Congress Control Number: 2013944538

Printed on acid-free paper

Springer is part of Springer Science+Business Media (www.springer.com)

Preface

It is very necessary to protect groundwater from contamination in order to ensure its availability for future generations. There is an increasing need for groundwater remediation because of human population expansion and on-going growth in industrial and agricultural development. Among various sources of groundwater contamination, nitrate is considered as the most ubiquitous chemical contaminant.

In view of nitrate contamination found in many aquifers of the world, a number of physical, chemical, or biological in situ remediation approaches have been reported over the past few years which can be applied by means of permeable reactive barrier (PRB) (See Chap. 1). Nevertheless, there remains a need to develop innovative and especially cost-effective remediation approaches. We have proposed a novel heterotrophic-autotrophic denitrification (HAD) approach which is supported by granulated spongy iron, pine bark, and mixed bacteria (See Chap. 2). The HAD involves biological deoxygenation, chemical reduction of nitrate and dissolved oxygen, heterotrophic denitrification and autotrophic denitrification. We have proposed two HAD PRBs as well and provided clear descriptions of their denitrification capacities (See Chap. 3). We have attempted to provide the reader with bacterial community structure and phylogenetic analysis (See Chap. 4).

We are extremely grateful to Erping Bi and Neil Buchanan for their useful suggestions and improving the English writing. We thank Jessica Hall, Michael Taylor, Natalie Bolton, Ryan Cheng, Aifang Jin, Xiangyu Guan, Jian Chen, Xiaopeng Qin, Jisheng Xu, Shengpin Li, Yingzhao Yang, and Hui Zhang for their helpful discussions and providing materials, labor, and moral support. We also warmly thank Nicholas White and Raj Indela for their technical assistance. We express thanks to Lisa Fan for her invaluable assistance in the preparation of this monograph. This monograph is financially supported jointly by National Program of Control and Treatment of Water Pollution (2009ZX07424-002-002), the project from the China Geological Survey (1212011121171), and Beijing Excellent Talents Program (2012D001055000001).

This monograph serves as a valuable resource to engineers, researchers, teachers, and students specialized in Environmental Science and Engineering, Hydrogeology and Engineering Geology, and Molecular Biology.

Beijing, China, May 2013

Fei Liu
Guoxin Huang

Contents

Chapter 1
General Introduction

Abstract High concentration of nitrate in drinking water is thought to be related to methemoglobinemia, cancers and even death. Due to the increasing anthropogenic activities, nitrate in groundwater is increasing in many areas of the world. Nitrate contamination is caused by nitrogenous fertilizers, livestock manures, agricultural irrigation, etc. This study overviewed the latest developments in nitrate *in situ* remediation and summarized advantages and disadvantages of each remediation approach. Currently physical adsorption (PA), biological denitrification and chemical reduction (CR) are the three approaches receiving considerable attention. Nitrate adsorbents in PA will ultimately get to the state of saturation due to adsorbed nitrate and its competing anions. BD is divided into heterotrophic denitrification (HD) and autotrophic denitrification (AD). A large number of liquid, solid and gas organic carbons in HD have been evaluated. For AD, hydrogenotrophic denitrification can be sustained by zero-valent iron (ZVI) which produces cathodic hydrogen. Low solubility of reduced sulfur species, sulfate production and biomass yield limit the applicability of sulfur autotrophic denitrification. The main disadvantage of ZVI-based CR is the release of ammonium under acidic conditions. More recently, a heterotrophic-autotrophic denitrification (HAD) approach has shown encouraging results. PA, cellulose-based HD, ZVI-based CR and AD, and their combined approaches can be applied by means of permeable reactive barrier (PRB). BD PRBs and ZVI PRBs have been successfully applied.

Keywords Nitrate · Groundwater · *In situ* remediation · Physical adsorption (PA) · Biological denitrification (BD) · Chemical reduction (CR) · Heterotrophic denitrification (HD) · Autotrophic denitrification (AD) · Zero-valent iron (ZVI) · Permeable reactive barrier (PRB)

Abbreviations

AD	Autotrophic denitrification
AMO-D	Aerobic methane-oxidation coupled to denitrification
ANMO-D	Anaerobic methane-oxidation coupled to denitrification
BATs	Best available technologies

F. Liu et al., *Study on Heterotrophic-Autotrophic Denitrification Permeable Reactive Barriers (HAD PRBs) for In Situ Groundwater Remediation*, SpringerBriefs in Water Science and Technology, DOI: 10.1007/978-3-642-38154-6_1, © The Author(s) 2014

BD	Biological denitrification
BET	Brunauer-Emmett-Teller
CR	Chemical reduction
DOC	Dissolved organic carbon
HAD	Heterotrophic-autotrophic denitrification
HD	Heterotrophic denitrification
HDTMA	Hexadecyltrimethyl ammonium bromide
HEPES	N-[2-hydroxyethyl]piperazine-N'-[2-ethanesulfonic acid]
HRT	Hydraulic retention time
HT	Hydrotalcite-type
ICs	Inert carriers
LOCSs	Liquid organic carbon sources
MOPS	3-(N-morpholino)propanesulfonic acid
NOCs	N-nitroso compounds
NTU	Nephelometric turbidity unites
NZVI	Nanoscale ZVI
PA	Physical adsorption
PCL	ε-caprolactone
PHB	3-hydroxybutyrate
PRB	Permeable reactive barrier
SOCSs	Solid organic carbon sources
TKN	Total kjeldahl nitrogen
VFAs	Volatile fatty acids
WHO	World health organization
ZVA	Zero-valent aluminium
ZVI	Zero-valent iron

1.1 Research Background

1.1.1 Nitrate Contamination of Groundwater

Groundwater is by far the largest freshwater resource on Earth other than water stored as ice (Bovolo et al. 2009). Groundwater is particularly important for drinking water supply and other domestic use as well as irrigation. However, groundwater is seriously contaminated by nitrate all over the world.

In Oceania, high nitrate-nitrogen (NO_3-N) concentrations (15–54 mg/L) have been observed in Northern Territory, Australia (Salvestrin and Hagare 2009). In Europe, NO_3-N contents in groundwater in farming areas increased from 5.4 mg/L during 1961–1965 to 9.7 mg/L in 1986 in Czechoslovakia (Beneš et al. 1989). NO_3-N concentrations in the northern coastal aquifers of Peloponnese were higher than 11.3 mg/L due to leaching of nitrogen fertilizer residuals in Greece

(Angelopoulos et al. 2009). In the period of 2000–2003, 17 % of monitoring stations (average values) had NO_3-N concentrations above 50 mg/L, 7 % stations were in a range of 40–50 mg/L, and 15 % stations were in a range of 25–40 mg/L in European Union (Angelopoulos et al. 2009). In South America, NO_3-N concentrations greater than the accepted level of 10 mg/L for safe drinking water were present in 36 % of the sampled wells in the Upper Pantanoso Stream Basin in Argentina (Costa et al. 2002). In North America, there are approximately 300–400 thousands of nitrate contaminated sites in the USA (Yang and Lee 2005). The more than 2.02×10^5 ha in the central Platte region are the largest areal expanse of nitrate contaminated groundwater in Nebraska, USA, and NO_3-N concentrations in this groundwater have increased at rates of 0.4–1.0 mg/L per year (Spalding and Exner 1993). In Africa, groundwater in a shallow lateritic aquifer showed NO_3-N enrichment ranging between 4.5 and 22.6 mg/L around the city of Sokoto, Nigeria (Bijay-Singh et al. 1995). In Asia, during the past 2 decades, NO_3-N concentrations in groundwater have increased steadily and reached or even exceeded the accepted level (10 mg/L) for drinking water in Japan (Kumazawa 2002). A survey of wells (n = 1,060) undertaken in all the 13 regions showed that the average NO_3-N concentrations greater than 10 mg/L were determined in Jizan, Asir, Qassim and Hail in Saudi Arabia (Alabdula'aly et al. 2010). NO_3-N contamination (≥ 20 mg/L) took place in the County Seat areas of Quzhou County, China (Hu et al. 2005) and its spatial distribution in the shallow groundwater was highly variable (Zhu and Chen 2002).

Apparently, nitrate contamination of groundwater has been found in each continent except Antarctica, so nitrate can be considered to be the most ubiquitous chemical contaminant in groundwater and its concentration have a tendency to gradually increase.

1.1.2 Sources of Groundwater Nitrate

Nitrite and ammonium will be transformed into nitrate when encountering oxygen in soils and groundwater. Nitrate itself is both soluble and mobile (Nolan 2001). These cause nitrate to be prone to leaching through soils with infiltrating water to groundwater. Groundwater nitrate contamination can be resulted from the following sources/mechanisms:

1. Nitrogenous fertilizers

Nitrogenous fertilizers are considered to be one of the principal sources of nitrate in groundwater. In the USA, some 1.05×10^7 t of nitrogen in fertilizers is applied each year to cropland and pastures (Hudak 2000). In China, the average nitrogenous fertilizer application is currently over 200 kg N/ha (similar to the Western Europe). In some areas in northern China where the NO_3-N concentration in groundwater can be as high as 67.7 mg/L, the nitrogenous fertilizer is applied in large quantities (500–1,900 kg N/ha), but the percentage of applied nitrogen taken up by crops is below 40 % (Zhang et al. 1996). Furthermore, fertilizer applications are expected to double or even triple within the next 30 years in China (Zhang et al. 1996).

2. Livestock manures

Livestock manures are another principal source of nitrate in groundwater. In the USA, animal manures contain 5.9×10^6 t of nitrogen (Hudak 2000). A study by Rao (2006) found that there was an association between high nitrate concentrations in wells and animal wastes at cattlesheds and sites close to animal trading markets.

3. Agricultural irrigation

Agricultural irrigation using domestic wastewater, industrial wastewater, reclaimed water and polluted surface water contributes to groundwater nitrate contamination. For example, given the average NO_3-N of 4.0 mg/L in the Yellow River, China, about 2.42×10^6 kg of NO_3-N pours annually into agricultural fields, rendering a maxmium NO_3-N of 22.8 mg/L in groundwater in some areas in 2003 (Chen et al. 2007).

4. Septic tanks, cesspools and pit latrines

Septic tanks, cesspools and pit latrines are great threats to groundwater quality. Septic tanks are a common practice in developing and developed countries. Human excreta contains about 5 kg N/cap a. The concentrations of NO_3-N and ammonium-nitrogen (NH_4-N) in effluent from a typical septic tank are <1 mg/L and 20–55 mg/L respectively. In the central eastern parts of Eskişehir, Turkey, which coincided with the densely populated and/or unsewered parts of the city, septic tanks played an important role in nitrate contamination (Kaçaroğlu and Günay 1997). In Amman, Jordan, 8×10^6 m^3/a from cesspool leakage seeped into groundwater (Salameh et al. 2002).

5. Contaminated land

Contaminated land, such as abandoned landfills, contributes a significant quantity of nitrogen to groundwater. In the past decades, landfill has led to serious groundwater contamination due to improper management (Dong et al. 2009). Typical concentrations of NO_3-N, nitrite-nitrogen (NO_2-N) and NH_4-N in landfill leachate were 0–1,250, 0–9.8 and 1.5 mg/L, respectively (Wakida and Lerner 2005).

6. River-aquifer interaction

It's rather common that an aquifer is contaminated by a river which receives raw or treated water and simultaneously infiltrates water to aquifers. In Turkey, the partial recharge of the groundwater in the Eskişehir Plain is by infiltration from the Pousuk River and its tributaries, into which municipal and industrial wastewater was discharged (Kaçaroğlu and Günay 1997). In a study, the average NO_3-N concentration (9.0 mg/L) in the groundwater was higher than that (4.1–6.7 mg/L) in the Porsuk River (Kaçaroğlu and Günay 1997). This indicated that the Porsuk River was a source of nitrate contamination in the groundwater, but it was not the only source. Riverbank infiltration can be enhanced by pumping stations. The infiltration amount increased from 0.61×10^4 to $(6.97–7.46) \times 10^4$ m^3/d under heavy pumping conditions at a pumping station in Changchun, China.

7. Atmospheric nitrogen deposition

The emission of nitrogen to atmosphere can be in two forms: nitrous oxides (NO$_x$) and ammonia (NH$_3$). The former is mainly from car engines and industry; the latter is mostly generated by agriculture and intensive feedstock rearing (Wakida and Lerner 2005). Both forms can later be deposited via dry and wet deposition. According to Boumans et al. (2004), about 35 % of 54 kg/ha a atmospheric N deposition was leachated to the upper groundwater, resulting in a NO$_3$-N concentration of about 6.8 mg/L. In the USA, approximately 3.2×10^6 t/a atmospheric N deposition was leachated to local groundwater (Steindorf et al. 1994).

1.1.3 Health Risks to Humans

Excessive levels of nitrate and nitrite in drinking water can cause serious illness and sometimes death. Public health concerns arise from the potential bacterial reduction of nitrate to nitrite. In a human body, about 25 % of ingested nitrate is recirculated in saliva, of which up to 20 % is converted to nitrite in the mouth (WHO 2008). The remaining nitrate is intaken into the stomach. It can be reduced significantly to nitrite by bacteria in the stomach under some circumstances such as a low gastric acidity or gastrointestinal infections (including individuals using antacids, particularly those that block acid secretion, and potentially bottle-fed infants under 6 months of age) (WHO 2008). This conversion can also take place at other sites including the distal small intestine and the colon (Ward et al. 2005). The reaction of nitrite with haemoglobin in red blood cells forms methemoglobin, which binds oxygen tightly and does not release it. As a result, oxygen transport in the body is blocked (Elmidaoui et al. 2002; Luk and Au-Yeung 2002; WHO 2008). High levels of methemoglobin (>10 %) can give rise to methemoglobinemia, referred to as blue-baby syndrome, whose symptoms include shortness of breath and blueness of the skin. During the time period of 1941–1949, 114 cases of infant methemoglobinemia were reported in Minnesota, including 14 deaths during a 30-month period. During 1945–1970, 2,000 cases with a fatal rate of 8 % were reported all over the world (Fan and Steinberg 1996).

Nitrate and nitrite are precursors in the endogenous formation of N-nitroso compounds (NOCs). A number of NOCs can cause hypertension (Gao et al. 2003), cancers (Deng 2000; Nolan and Hitt 2006), malformation and mutation (Liu et al. 2009). Cancer incidence was analyzed in a cohort of 21,977 Iowa women.

The analysis results indicated that there were positive trends between nitrate levels and risk of bladder and ovarian cancers (Weyer et al. 2001).

In addition, there was a possible causal relation between ingesting nitrate contaminated water and spontaneous abortion (Ruckart et al. 2008). This relation was demonstrated by an investigation involving women with spontaneous abortions in LaGrange County, Indiana, USA.

To avoid the risks mentioned above, a large number of countries, areas and organizations established stringent stands for consumption water, particularly drinking water.

1.2 In Situ Remediation Approaches for Nitrate Contaminated Groundwater

Reverse osmosis, ion exchange and electrodialysis are considered as the best available technologies (BATs) to *ex situ* remove nitrate from water. However, BATs lead to the technological and constructional complexity in their *in situ* application. Another problem with BATs is a large excess of nitrate concentrated waste streams produced by the regeneration of membrane (or resin), which have the risk of secondary pollution. Actually, BATs do not reduce nitrate into nontoxic N-containing substances, but just transfer it.

In situ remediation is more advantageous compared to *ex situ* treatment because it utilizes a natural aquifer as a physiochemical and/or biochemical reactor and a filter for pollutant and bioproduct removal (Haugen et al. 2002). Other advantages include low equipment complexity, low energy consumption, low interference with surface activity, low long-term liability and stable groundwater temperature.

Currently nitrate treatment approaches which can be applicable *in situ* are: (1) physical adsorption (PA); (2) biological denitrification (BD); and (3) chemical reduction (CR).

1.2.1 Physical Adsorption

Generally speaking, adsorption on a solid support is the process of collecting soluble substances from water solution to the absorbent particles. Materials which are reported to remove nitrate are summarized here:

- Activated carbon (virgin activated carbon, activated carbon impregnated with $MgCl_2$, activated carbon modified by $ZnCl_2$);
- Bamboo powder charcoal (from the residual of Moso bamboo manufacturing);
- Chitosan (cross-linked chitosan gel beads; chitosan extracted from the waste of shrimps);

- Ion exchange resins (Amberlite IRA 410, Duolite A 196, Amberlite IRA 996, Purolite A 520);
- Organoclays-organic surfactants modified clay minerals;
- Phosphoric acid dibutyl ester type;
- Sepiolite;
- Surfactant modified zeolites.

Adsorption performances of the above materials are studied by numerous researchers. Activated carbon impregnated with $MgCl_2$ or $ZnCl_2$, which has a high pore volume and a large Brunauer-Emmett-Teller (BET) surface area (Demiral and Gündüzoğlu 2010), is more efficient than virgin activated carbon. For example, the maximum nitrate removal efficiency was 74 % for activated carbon modified by $MgCl_2$, while it was only 8.8 % for virgin activated carbon (Rezaee et al. 2010). Bamboo powder charcoal also exhibits higher nitrate removal efficiency than virgin activated carbon (Mizuta et al. 2004). Maximum adsorption capacities for nitrate on chitosan were 8.03 mg/g (from Langmuir isotherm) and 23.85 mg/g (from Freundlich isotherm); while its experimental value was 19 mg/g (Menkouchi Sahli et al. 2008). For Amberlite IRA 410, the nitrate removal could be best described by the pseudo second order and intraparticle diffusion models (Chabani and Bensmaili 2005). For Duolite A 196, Amberlite IRA 996 and Purolite A 520, their selectivity is better for nitrate than for sulphate (Boumediene and Achour 2004). Therefore they have potential application prospects in groundwater treatment when groundwater presents a strong content of sulphate. All untreated clays, including bentonite, kaolinite and halloysite, have poor adsorption capacities for nitrates, whereas these clays modified with surfactant hexadecyltrimethylammonium bromide (HDTMA) greatly improves their removal capacities (Xi et al. 2010). Similarly, HDTMA modified zeolite has been shown to be useful for nitrate removal (Guan et al. 2010). In comparison to virgin clay, zeolite has better mechanistic strength and hydraulic permeability for *in situ* reactive barriers. The absorbent with a phosphoric acid dibutyl ester type active group is useful for HNO_3 or NO_3-N removal. The maximum amount of HNO_3 adsorbed is nearly equal to the active group content (Sato et al. 1995). Sepiolite activated by HCl is more effective for nitrate removal compared with virgin sepiolite. The equilibrium time was 30 and 5 min for sepiolite (200–170 mesh) and sepiolite activated by HCl (200–170 mesh), respectively (Öztürk and Bektaş 2004).

All the above mentioned materials (natural, synthetic and modified) have large specific surface areas contributing to the capacity for nitrate extraction from aqueous solution. However, without nitrate degradation, these adsorbents will become saturated and lose the removal capacity. The adsorbent regeneration can be a technical problem and costly. Moreover, presence of numerous competing anions (Cl^-, SO_4^{2-}, etc.) in groundwater can exacerbate adsorbent saturation problem.

1.2.2 Biological Denitrification

BD is central to the nitrogen cycle with respect to the subsurface groundwater environment (Rivett et al. 2008). BD can offer *in situ* treatment process for contaminated groundwater thanks to high specificity of denitrifying bacteria, low cost and high denitrification efficiency (Wang et al. 2009). BD is an anoxic or anaerobic process in which nitrate is reduced into nitrous oxides and subsequently to harmless nitrogen by means of the action of denitrifying bacteria (Eq. 1.1) (Fernández-Nava et al. 2010).

$$NO_3^- \rightarrow NO_2^- \rightarrow NO(g) \rightarrow N_2O(g) \rightarrow N_2(g) \tag{1.1}$$

These transformations in Eq. 1.1 involve oxidized nitrogen compound dissimilatory reduction in which nitrate and nitrite are used as ending electron acceptors (Moreno et al. 2005).

With regard to carbon sources, BD is divided into two ways: heterotrophic denitrification (HD) and autotrophic denitrification (AD).

1.2.2.1 Heterotrophic Denitrification

HD processes are the most studied by numerous researchers and most widely applied in the field (Soares et al. 2000). The majority of microbial denitrification processes depend on heterotrophic denitrifiers, which require organic carbon substrates as their electron donors and energy sources and tend to use them as sources of cellular carbon. HD does not occur in a deep aquifer due to insufficient organic carbons and electron donors (Starr and Gillham 1993). Intrinsic nitrate degradation (by denitrification) may become very slow and inadequate to protect groundwater (Devlin et al. 2000). Therefore external organic carbon sources (in particular readily biodegradable ones) have to be supplied for bacterial growth, enrichment and respiration. Generally, organic carbon sources are classified into three groups of solid, liquid and gas carbonaceous substances.

1. Liquid organic carbon

Many typical liquid organic carbon sources (LOCSs) including volatile fatty acids (VFAs), acetate, formate, starch, sucrose, ethanol, methanol, and vegetable oils have been widely studied.

The utilization pattern of VFAs in a BD process was investigated. The average specific denitrification rate for natural VFAs (0.0111 g NO_3-N/g VSS d) was close to the mixture of synthetic VFAs (0.0134 g NO_3-N/g VSS d); and the average specific carbon consumption rate for natural VFAs (0.0252 g VFA-C g/g VSS d) was similar to that for synthetic VFAs (0.0248 g VFA-C g/g VSS d) (Elefsiniotis and Wareham 2007).

An off-line municipal well was chosen to demonstrate the practicality of *in situ* biodenitrification with acetate at Wahoo, Nebraska, USA. The *in situ* denitrification process was sustained for 3 months without evidence of clogging (Khan and Spalding 2004). Pulsed injection of acetate resulted in more evenly distributed biomass profiles (Peyton 1996).

Sodium formate was added to a sand and gravel aquifer on Cape Cod, MA, USA to test whether formate could act as an electron donor for *in situ* denitrification (Smith et al. 2001). The results showed that the nitrate concentration decreased with time. Given the capacity of denitrifiers to grow aerobically, it was likely that formate could be used to remove both oxygen and then nitrate.

Nontoxic soluble starch and a facultative psychrophilic denitrifier (strain 47) were utilized for on-site groundwater denitrification (Kim et al. 2002). The results indicated that the nitrogen removal efficiency of 99.5 % was obtained at a hydraulic retention time (HRT) of 1 h with a C/N weight ratio of 2.58.

The effectiveness of sucrose, ethanol and methanol for nitrate removal was studied. Ethanol and methanol with a higher density of denitrifying bacteria increased denitrification activity compared with sucrose, and sucrose produced a greater biomass, causing clogging (Gómez et al. 2000).

The ability of a vegetable oil-based barrier to remove nitrate was examined. During the 30-week study, 39 % NO_3-N was removed from the groundwater at an initial NO_3-N of 20 mg/L (Hunter 2001). In addition, vegetable oils spreading uniformly over soil may prevent localized accumulation of biomass (Hunter 2001).

LOCSs have shown some advantages such as high effectiveness, widespread availability, easy handling and low specific cost (e.g. €2.0–4.0/kg NO_3-N for methanol; €2.4/kg NO_3-N for ethanol) (Boley et al. 2000). However, their use requires reservoirs for the substrate solution and the continued (or intermediate) operation of pumping systems. The pump(s), reservoirs and associated plumbing add to the operational complexity of these systems (Peyton 1996).

2. Solid organic carbon

Solid organic carbon sources (SOCSs) such as cotton, wood chips, sawdust and crab-shell chitin serve as not only carbon sources and energy substrates but also the physical support for microorganisms (Fig. 1.1). Recently, various studies have been conducted to evaluate potential use of SOCSs which mainly include cellulose-based substrates and biodegradable polymers.

(a) cellulose-based substrates

Cellulose is the most abundant renewable resource in the world because it is a very basic component of all plant materials. Cellulose consists of linear glucose polymers with hydrogen bonding between hydroxyl groups of neighbouring parallel chains (Volokita et al. 1996a).

Cotton is the purest form of native cellulose with only a small percentage of impurities mostly in the form of wax, pectin and protein residues (Volokita et al. 1996b). Cotton is the best cellulose-based substrate with the highest specific external surface to support HD (Volokita et al. 1996b). Della Rocca et al. (2005) reported that the use of cotton exhibited very good nitrate removal performance

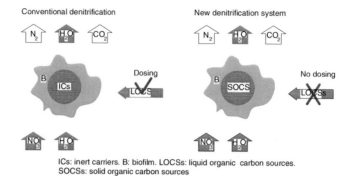

ICs: inert carriers. B: biofilm. LOCSs: liquid organic carbon sources.
SOCSs: solid organic carbon sources

Fig. 1.1 Solid organic carbon-supported biological denitrification (BD) (modified from Boley et al. 2000)

(percent removal of >90 %), and no significant nitrite accumulation occured in the denitrified water.

Wheat straw, sawdust and wood chips are effective for enhancing *in situ* bio-denitrification. High denitrification rates with wheat straw were observed during the first week of operation, but then the efficiency declined (Soares and Abeliovich 1998), which was related to the release of dissolved organic carbon (DOC) (Aslan and Turkman 2003). This implies that wheat straw can not steadily provide organic carbon. Sawdust and wood chips do not increase effluent total kjeldahl nitrogen (TKN) (≤ 0.7 mg/L as N) and turbidity [≤ 2.4 nephelometric turbidity unites (NTU)] (Kim et al. 2003). In a BD with wood chips, in conjunction with the decrease in nitrate, SO_4^- decreased and HCO_3^-, Fe (II) and Mn (II) increased in the effluent water (Blowes et al. 1994). Although wood chips lost 16.2 % of its mass during the 140-day operation (Saliling et al. 2007), they could provide steady denitrification rates (Greenan et al. 2006). Additionally, the addition of soybean oil to wood chips can significantly increase denitrification efficiency over wood chips alone (Greenan et al. 2006). Newspaper is an excellent solid phase electron donor substrate (Kim et al. 2003). Complete removal of NO_3-N (22.6 mg/L) was achieved without NO_2-N accumulation by newspaper, and the treated water contained low DOC (4–10 mg/L) (Volokita et al. 1996a). Unfortunately, pine bark as a solid carbon source has not been reported.

(b) Biodegradable polymers

Biodegradable polymers can be used as alternative insoluble carbon sources and biofilm carriers for HD. 3-hydroxybutyrate (PHB, $[C_4H_6O_2]_n$, model BIOPOL D400 GN), ε-caprolactone (PCL, $[C_6H_{10}O_2]_n$, model TONE P 787) and Bionolle ($[C_6H_8O_4]_n$, model # 6010) obtained excellent nitrate removal performance (Boley et al. 2000). However, PHB and PCL lead to very expensive approaches with €6.6–8.9/kg NO_3-N and €21.0–37.2/kg NO_3-N respectively. Chitin is the second most abundant biopolymer in nature. Chitin degradation proceeds mainly through fermentation and hydrolysis. Chitin fermentation produces VFAs and hydrogen

(Brennan et al. 2006a, b). Chitin hydrolysis results in the production of N-acetylglucosamine monomers. Nevertheless, the monomers will release nitrogen (Brennan et al. 2006b). Robinson-Lora and Brennan (2009) reported crab-shell chitin was an attractive electron donor and carbon source for groundwater biodenitrification. Its denitrification rates were 2.4 ± 0.2 mg N/L d in batch tests, but rapid degradation of protein caused an initial high release of carbon and ammonium (Robinson-Lora and Brennan 2009).

3. Gas organic carbon

Methane is a potentially inexpensive, widely available external substrate for groundwater biodenitrification. The infiltered gas will spread fast in aquifers, reduce the risk of clogging near infiltration wells and increase the microbial activity. However, methane solubility should deserve a closer attention because it is poorly water soluble (Henry constant, 1.4×10^{-3} mol/kg/atm; water diffusivity, 18.8×10^{-10} m^2/s).

Methane has been used in BD under two completely different environmental conditions: (1) an aerobic condition; (2) an anaerobic (or anoxic) condition.

Under an aerobic condition, aerobic methane-oxidation coupled to denitrification (AMO-D) is accomplished by aerobic methanotrophs that oxide methane, consume oxygen and simultaneously release soluble simple organic intermediates. The released organic intermediates are subsequently used by coexisting heterotrophic denitrifiers as electron donors at anaerobic conditions created by methanotrophic bacteria in groundwater (Rajapakse and Scutt 1999; Knowles 2005; Modin et al. 2007). It is obvious that AMO-D leads to an "indirect" denitrification. It is worth mentioning that the soluble organics may contain methanol (Eisentraeger et al. 2001; Knowles 2005), acetate (Eisentraeger et al. 2001), proteins (Eisentraeger et al. 2001). AMO-D can be induced at a low temperature of 10 °C (Eisentraeger et al. 2001). Nevertheless, the process of AMO-D provoked an increase of biomass and high wasteful methane oxidation if a successful process control is not in place (Thalasso et al. 1997). Furthermore, the off-gas containing a mixture of O_2 and CH_4 made CH_4 unusable for fuel and carried explosion risks.

Under an anaerobic condition, anaerobic methane-oxidation coupled to denitrification (ANMO-D) is mediated by an association of archaeons and bacteria that oxidize methane to carbon dioxide coupled to denitrification (Eqs. 1.2, and 1.3) (Raghoebarsing et al. 2006; Modin et al. 2007). Obviously, ANMO-D leads to a "direct" denitrification.

$$CH_4 + 8NO_3^- + 8H^+ \rightarrow 5CO_2 + 4N_2 + 14H_2O \tag{1.2}$$

$$(\Delta G^{0\prime} = -765 \, \text{kJ mol}^{-1} \, CH_4)$$

$$3CH_4 + 8NO_2^- + 8H^+ \rightarrow 3CO_2 + 4N_2 + 10H_2O \tag{1.3}$$

$$(\Delta G^{0\prime} = -928 \, \text{kJ mol}^{-1} \, CH_4)$$

However, the mechanisms of ANMO-D microorganisms are still unclear.

1.2.2.2 Autotrophic Denitrification

Complex organic substances can not be utilized by autotrophic denitrifying bacteria as oxidizable substrates. In contrast, carbon dioxide and bicarbonate are utilized as carbon sources for microbial cell synthesis under autotrophic growth conditions. Some bacteria can be capable of utilizing hydrogen as well as various reduced sulfur compounds (S^0, H_2S, S^{2-}, $S_2O_3^{2-}$, $S_4O_6^{2-}$, SO_3^{2-}, etc.) as electron donors and energy sources for microbial metabolic chain (Matějů et al. 1992). The potential advantages of autotrophic over heterotrophic denitrification are: low biomass buildup (biofouling); reduction of clogging; evasion of poisoning effect of some organic carbons; avoidance of organic carbon contamination of treated water; and easier post-treatment (van Rijn et al. 2006; Ghafari et al. 2008).

With regard to electron donors, AD is divided into two ways: Hydrogenotrophic denitrification and sulfur autotrophic denitrification.

1. Hydrogenotrophic denitrification

Hydrogen gas is an ideal substrate for BD in that it is completely harmless to human health and naturally clean, and no further post-treatments are required to remove either excess residues or its derivatives in addition to the general advantages of AD.

The pathways for hydrogenotrophic denitrification are given in the following reactions (Eqs. 1.4–1.8, and 1.9) (Chang et al. 1999; Lee and Rittmann 2002; Lee et al. 2010; Karanasios et al. 2010).

$$NO_3^- + H_2 \rightarrow NO_2^- + H_2O \tag{1.4}$$

$$NO_2^- + 0.5H_2 + H^+ \rightarrow NO + H_2O \tag{1.5}$$

$$2NO + H_2 \rightarrow N_2O + H_2O \tag{1.6}$$

$$N_2O + H_2 \rightarrow N_2 + H_2O \tag{1.7}$$

$$2NO_3^- + 5H_2 \rightarrow N_2 + 4H_2O + 2OH^- \text{ (overall reaction)} \tag{1.8}$$

$$2NO_3^- + 2H^+ + 5H_2 \rightarrow N_2 + 4H_2O \text{ (overall reaction)} \tag{1.9}$$

$$H_2 + 0.35NO_3^- + 0.35H^+ + 0.052CO_2 \rightarrow 0.17N_2 + 1.1H_2O + 0.010C_5H_7O_2N \tag{1.10}$$

Based on Eq. 1.10 (Ergas and Reuss 2001), the cell yield is approximately 0.24 g cells/g NO_3-N. This value is considerably lower than the 0.6–0.9 g cells/g NO_3-N typically reported for HD.

There are some drawbacks with sparging H_2 into aquifers. Due to its low solubility (1.6 mg/L at 20 °C) in water, the addition of hydrogen to groundwater is not straightforward. Considering the stoichiometry of $\sim$6:1 (H_2:NO_3^-) for hydrogenotrophic denitrification (Smith et al. 1994), if groundwater contains

Fig. 1.2 Zero-valent iron (ZVI)-supported hydrogenotrophic denitrification (modified from Till et al. 1998)

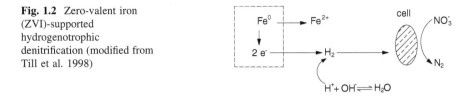

10 mg/L NO_3-N, only ~ 20 % of NO_3-N can be reduced, even though the water is fully saturated with hydrogen. Therefore, effective approaches are needed to obtain sufficient substrate delivery and mass transfer from gas phase to liquid phase (Smith et al. 2001). Additionally, H_2 gas forms high flammable and explosive mixtures with O_2 gas during the operation, transformation and storage. Its explosion and safety concerns have prevented widespread acceptance of the hydrogenotrophic denitrification.

Recently, it has been demonstrated that hydrogenotrophic denitrification can be sustained by various zero-valent iron (ZVI) species (Eq. 1.11; Fig. 1.2) which can produce cathodic hydrogen during anaerobic ZVI corrosion by water (Till et al. 1998). Cathodic hydrogen overcomes the limitation associated with hydrogen delivery. The mechanism of anaerobic ZVI corrosion is cathodic depolarization, in which electrons from ZVI and H^+ from water produce molecular hydrogen (H_2), as given in Eq. 1.12 (Daniels et al. 1987). The bacteria in water accelerate the anodic dissolution of ZVI by using hydrogen through their hydrogenase enzymes (Daniels et al. 1987).

$$5Fe^0 + 2NO_3^- + 6H_2O \rightarrow 5Fe^{2+} + N_2 + 12OH^- \qquad (1.11)$$

$$(\Delta G^{0'} = -1,147\,kJ)$$

$$Fe^0 + 2H_2O \rightarrow H_2 + Fe^{2+} + 2OH^- \qquad (1.12)$$

Fe^{2+} could also act as another electron donor for nitrate removal in the nanoscale ZVI (NZVI) mediated microbial system, which might be more applicable for *in situ* remediation than other alternative technologies (Shin and Cha 2008).

2. Sulfur autotrophic denitrification

AD supported with reduced sulfurs owns extra advantages in addition to the general advantages of AD. Elemental sulfur is less expensive compared to ethanol, methanol, etc. because it is a by-product of oil processing. S-autotrophic denitrification can take place under aerobic conditions (Zhang and Lampe 1999), which eliminates the requirement to deoxygenate water. Nevertheless, there are some drawbacks with S-autotrophic denitrification: low solubility of reduced sulfurs, use of limestone for pH adjustment and undesirable by-products sulphates (Karanasios et al. 2010). Sulfate production and biomass yield were usually higher under aerobic conditions than anaerobic ones (Zhang and Lampe 1999). These drawbacks limit its *in situ* applicability.

S-autotrophic denitrification has been studied for groundwater treatment. Zhang and Lampe (1999) pointed out that the optimal sulfur: limestone ratio was 3:1 (v:v) when limestone served as a source of inorganic carbon and pH buffering agent. Moon et al. (2006) reported TCE of 80 mg/L, Zn of $\geq$0.5 mg/L and Cu of $\geq$0.5 mg/L markedly inhibited S-autotrophic denitrification. Moon et al. (2008) reported phosphate was crucial to the denitrification activity. The presence of sulfur compounds and nitrate is of prime importance for biomass yield and operation. Based on Eqs. 1.13, and 1.14 (Sierra-Alvarez et al. 2007) and Eqs. 1.15, 1.16, and 1.17 involving cell yields (Campos et al. 2008), NO_3-N is reduced to N_2 and hydrogen ions are simultaneously produced, which means alkalinity is consumed.

$$S^0 + 1.2NO_3^- + 0.4H_2O \rightarrow SO_4^{2-} + 0.6N_2 + 0.8H^+ (\Delta G^{0'} = -547.6\,kJ) \quad (1.13)$$

$$S_2O_3^{2-} + 1.6NO_3^- + 0.2H_2O \rightarrow SO_4^{2-} + 0.8N_2 + 0.4H^+ (\Delta G^{0'} = -765.7\,kJ) \tag{1.14}$$

$$1.10S^0 + NO_3^- + 0.76H_2O + 0.40CO_2 + 0.08NH_4^+$$
$$\rightarrow 1.10SO_4^{2-} + 0.50N_2 + 0.08C_5H_7O_2N + 1.28H^+ \tag{1.15}$$

$$0.421H_2S + 0.421HS^- + NO_3^- + 0.346CO_2 + 0.086HCO_3^- + 0.086NH_4^+$$
$$\rightarrow 0.842SO_4^{2-} + 0.500N_2 + 0.086\,C_5H_7O_2N + 0.434H_2O + 0.262H^+ \tag{1.16}$$

$$0.844S_2O_3^{2-} + NO_3^- + 0.347CO_2 + 0.086HCO_3^- + 0.086NH_4^+ + 0.434H_2O$$
$$\rightarrow 1.689SO_4^{2-} + 0.500N_2 + 0.086\,C_5H_7O_2N + 0.697H^+ \tag{1.17}$$

1.2.3 Chemical Reduction

CR of nitrate has been studied extensively using various substances, mainly including hydrogen gas, formic acid, zero-valent aluminium (ZVA) and ZVI (Prüsse and Vorlop 2001; Luk and Au-Yeung 2002; Huang and Zhang 2004).

With hydrogen gas or formic acid as a reductant in the presence of a bimetallic catalyst (such as Pd–Cu, Pd–Sn, Pd–In), nitrate may be converted to either nitrite as a toxic intermediate, or nitrogen as an ideal product, or ammonium as an undesired by-product (Fig. 1.3) (Prüsse and Vorlop 2001). These technologies are based on the catalytic hydrogenation of nitrate and nitrite. *In-situ* application potentiality of catalytic hydrogenation is very low. A hydrogen-based catalytic reaction needs a complex underground structure which includes the use of very expensive palladium-based catalysts and hydrogen injection devices (Della Rocca et al. 2007a). Moreover, the groundwater impurity could quickly inactive the effect of catalysts (Della Rocca et al. 2007a).

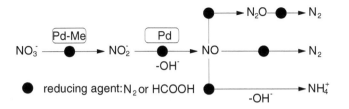

Fig. 1.3 Scheme of catalytic hydrogenation of nitrate by bimetallic catalysts (Prüsse and Vorlop 2001)

Nitrate reduction by ZVA may be described by Eq. 1.18 (Luk and Au-Yeung 2002).

$$3NO_3^- + 8Al + 18H_2O \rightarrow 3NH_3 + 8Al(OH)_3\downarrow + 3OH^- \qquad (1.18)$$

A maximum of 62 % NO_3-N removal was achieved using ZVA under the conditions of initial NO_3-N of 20 mg/L, aluminum of 300 mg/L, pH of 10.7, water temperature of 25 °C. On completion of the treatment, NO_3-N of 8.3 mg/L, NO_2-N of 0.26 mg/L and NH_4-N of 0.50 mg/L were measured, respectively, all within the maximum acceptable concentrations of the Canadian guidelines (Luk and Au-Yeung 2002).

In situ nitrate CR by ZVI is receiving considerable attention because ZVI is readily available at low cost and nontoxic, and its reduction process is a rapid reaction if the solution pH remains within an acidic range (Ruangchainikom et al. 2006; Ahn et al. 2008). Based on a literature survey, possible pathways for ZVI-based nitrate CR are summarized (Table 1.1).

The main disadvantage of ZVI-based CR is the release of ammonium as a major undesirable nitrogen product under acidic conditions (Table 1.1). Ammonium must be removed by post treatments, because it would be nitrified and transformed into nitrate again by nitrification bacteria in the presence of oxygen gas. However, NZVI performs better compared to other forms, because the end product was not ammonium but N_2 gas (Choe et al. 2000).

Solution pH influences the nitrogen end products (Table 1.1) and denitrification capacity of ZVI. ZVI could chemically treat nitrate rapidly at low pH of 2.0–5.0, but slowly at a neutral or weakly alkaline pH in an unbuffered solution (Cheng et al. 1997; Huang et al. 1998; Huang and Zhang 2004; Su and Puls 2004; Choe et al. 2004; Ahn et al. 2008). A pH buffer enhances nitrate reduction (Cheng et al. 1997; Ahn et al. 2008). As for NZVI, complete denitrification can be achieved without a pH buffer (Choe et al. 2000). Nitrate reduction by ZVI will result in an increase in solution pH which inhibits abiotic and biotic activities. The rise in pH is due to either anaerobic or aerobic iron corrosion thereby producing hydroxide ions (Eqs. 1.12, and 1.19) (Cheng et al. 1997; Su and Puls 2004; Kielemoes et al. 2000).

$$2Fe^0 + O_2 + 2H_2O \rightarrow 2Fe^{2+} + 4OH^- \qquad (1.19)$$

Table 1.1 Possible pathways for zero-valent iron (ZVI)-based nitrate chemical reduction (CR)

ZVI	Solution pH	Proposed pathway (s)	Reference
Iron powder (100–200 mesh)	0.1 M HEPES (N-[2-hydroxyethyl]piperazine-N'-[2-ethanesulfonic acid]) pH buffer; initial pH = 7	$10Fe^0 + 6NO_3^- + 3H_2O \rightarrow 5Fe_2O_3 + 3N_2 + 6OH^-$; $Fe^0 + NO_3^- + 2H^+ \rightarrow Fe^{2+} + H_2O + NO_2^-$	Siantar et al. (1996)
Iron powder (325 mesh)	0.05 M acetate/acetic acid buffer; initial pH = 5.0	$4Fe^0 + NO_3^- + 10H^+ + \rightarrow 4Fe^{2+} + NH_4^+ + 3H_2O$	Cheng et al. (1997)
Iron powder (6–10 μm)	No pH buffer; initial pH = 2	$Fe^0 + 2H_3O^+ \rightarrow H_2 + Fe^{2+} + 2H_2O$; $Fe^0 + NO_3^- + 2H_3O^+ \rightarrow Fe^{2+} + NO_2^- + 3H_2O$; $4Fe^0 + NO_3^- + 10H_3O^+ \rightarrow 4Fe^{2+} + NH_4^+ + 13H_2O$	Huang et al. (1998)
Nanoscale iron (1–100 nm)	No pH buffer; initial pH = ~7	$5Fe^0 + 2NO_3^- + 6H_2O \rightarrow 5Fe^{2+} + N_2 + 12OH^-$	Choe et al. (2000)
Iron powder (40 mesh)	50 mM MOPS (3-(N-morpholino)propanesulfonic acid) pH buffer; initial pH = 7	$4Fe^0 + NO_3^- + 10H^+ \rightarrow NH_4^+ + 3H_2O + 4Fe^{2+}$; $3Fe^0 + NO_2^- + 8H^+ \rightarrow NH_4^+ + 2H_2O + 3Fe^{2+}$	Alowitz and Scherer (2002)
Iron powder (100 mesh)	No pH buffer; initial pH = 2	$4Fe^0 + NO_3^- + 7H_2O \rightarrow 4Fe^{2+} + NH_4^+ + 10OH^-$	Choe et al. (2004)
Iron powder (100 mesh)	No pH buffer; initial pH = 3	$4Fe^0 + NO_3^- + 7H_2O \rightarrow 4Fe^{2+} + NH_4^+ + 10OH^-$	Choe et al. (2004)
Iron powder (100 mesh)	No pH buffer; initial pH = 6.2	$4Fe^0 + NO_3^- + 7H_2O \rightarrow 4Fe^{2+} + NH_4^+ + 10OH^-$	Choe et al. (2004)
Iron grains (~0.5 mm)	HCl pH buffer; initial pH = 1.95–2.05	$4Fe^0$ (coated with an iron oxide) $+ NO_3^- + 10H^+ \rightarrow 4Fe^{2+} + NH_4^+ + 3H_2O$	Huang and Zhang (2004)
Nanoscale iron (50–80 nm)	pH ≤ 3	$4Fe^0 + NO_3^- + 10H^+ \rightarrow 4Fe^{2+} + NH_4^+ + 3H_2O$ or $5Fe^0 + 2NO_3^- + 12H^+ \rightarrow 5Fe^{2+} + N_2 + 6H_2O$	Yang and Lee (2005)
Iron powder (10 μm) (+CO₂)	No pH buffer; initial pH ≈ 7	$4Fe^0 + NO_3^- + 10H^+ \rightarrow 4Fe^{2+} + NH_4^+ + 3H_2O$; $5Fe^0 + 2NO_3^- + 12H^+ \rightarrow 5Fe^{2+} + N_2 + 6H_2O$; $Fe^0 + NO_3^- + 2H^+ \rightarrow Fe^{2+} + NO_2^- + H_2O$	Ruangchainikom et al. (2006)
Peerless iron (>0.5 mm)	0.2 M NaAC-HAC pH buffer; initial pH = 3.81	$Fe^0 + H_2O + H^+ \rightarrow Fe^{2+} + H_2 + OH^-$; $2Fe^0 + O_2 + 2H_2O \rightarrow 2Fe^{2+} + 4OH^-$; $2Fe^0 + O_2 + 4H^+ \rightarrow 2Fe^{2+} + 2H_2O$; $4Fe^0 + NO_3^- + 10H^+ \rightarrow 4Fe^{2+} + NH_4^+ + 3H_2O$; $Fe^0 + NO_3^- + 2H^+ \rightarrow Fe^{2+} + H_2O + NO_2^-$; $3Fe^0 + NO_2^- + 8H^+ \rightarrow 3Fe^{2+} + NH_4^+ + 2H_2O$	Su and Puls (2007)
Nanoscale iron (~100 nm)	Initial pH = ~7	$4Fe^0 + NO_3^- + 10H^+ \rightarrow 4Fe^{2+} + NH_4^+ + 3H_2O$	Shin and Cha (2008)
Granulated iron (0.42 mm)	Initial pH = 2	$4Fe^0 + NO_3^- + 10H^+ \rightarrow 4Fe^{2+} + NH_4^+ + 3H_2O$	Rodriguez-Maroto et al. (2009)

Interferences with reactions by other substances can take place when ZVI is applied to treat nitrate contaminated groundwater. Some organic and inorganic ligands in soils, sediments and groundwater significantly reduce nitrate reduction rates. Phosphate and citrate definitely retard nitrate reduction by ZVI (Su and Puls 2004). Certain types of DOC have been shown to block the reactive sites on ZVI (ITRC 2005). Liao et al. (2003), for instance, found that the presence of propanol appeared to inhibit nitrate removal. H_2 gas, which is formed as a product of iron corrosion, might temporarily passivate iron surface and reduce porosity of ZVI (Thiruvenkatachari et al. 2008).

In summary, the drawbacks associated with nitrate reduction with ZVI are as follows: release of ammonium, requirement of low pH conditions and/or pH buffers, pH increase in absence of a pH buffer, slow removal efficiency at a neutral or weakly alkaline pH, and passivation of the ZVI surface.

1.2.4 Combined Approaches

Recently, several combined approaches of HD and AD are examined for nitrate removal from water. Liu et al. (2009) developed a combined two-step approach in which methanol was consumed by heterotrophic denitrifiers (Eq. 1.20), and then elemental sulfur was utilized by autotrophic denitrifiers (Eq. 1.21). As seen from Eqs. 1.20, and 1.21, pH can be better maintained in the treated water.

$$NO_3^- + 1.08CH_3OH + H^+ \rightarrow 0.065C_5H_7O_2N + 0.467N_2 + 0.76CO_2 + 2.44H_2O$$
$$(1.20)$$

$$1.11S^0 + 1.06NO_3^- + 0.785H_2O + 0.3CO_2$$
$$\rightarrow 1.11SO_4^{2-} + 0.5N_2 + 0.06C_5H_7O_2N + 1.16H^+ \qquad (1.21)$$

Batch tests were carried out to study the removal of nitrate in the single, binary, and ternary systems of cotton burr compost, ZVI and sediment (Su and Puls 2007). The results showed cotton burr compost alone removed nitrate at a faster denitrification rate than did cotton burr compost mixed with ZVI. This implied that ZVI in the ZVI/cotton burr compost system retarded nitrate removal. More recently, a novel heterotrophic-autotrophic denitrification (HAD) approach was proposed by Della Rocca et al. (2006, 2007b). Steel wool followed by cotton has been applied to support the combined HAD in continuous pilot-scale columns (Della Rocca et al. 2006, 2007b). The synergistic effects of HD and AD in these columns were encouraging. Nevertheless, the HAD process demonstrated some shortcoming such as high bacterial colony formation in the outlet, and nitrite accumulation and ammonium production at high inlet NO_3-N concentrations (Della Rocca et al. 2006).

1.3 Permeable Reactive Barriers for Nitrate Removal

PA, cellulose-based HD, ZVI-based CR and AD, and their combined approaches can be applied by means of permeable reactive barrier (PRB), which provides passive *in situ* remediation of nitrate in groundwater by promoting a variety of physical, chemical and biological reactions as groundwater flows through them. The main advantage of PRB is the passive nature of the treatment (Gavaskar 1999). That is, for the most part, its operation does not depend on any external labor or energy inputs (Gavaskar 1999). Once installed, the barrier takes advantage of groundwater flow (natural gradient) to bring nitrate in contact with reactive materials (Gavaskar 1999). Although PRB may involve substantial initial capital cost, particularly as depths increase, it is cost-effective in comparison with less-passive treatment technologies such as pump and treat because of lower operating and maintenance costs (Robertson et al. 2008).

1.3.1 Denitrification Reactive Media

Reactive media are the core of PRB application because they determine the operational performance and costs. The following important considerations should be made for the choice of denitrification reactive media.

- Reactivity—The reaction rate of nitrate with reactive media should be high so as to reduce the required HRT and therefore the size of PRB;
- Stability—Reactive media should not be soluble or depleted by reactivity, but should persist in the subsurface environment for a long enough period of time;
- Availability—Reactive media should be readily available at a reasonable medium–low cost;
- Hydraulic performance—The permeability of reactive media should be ten times greater than that of its surrounding aquifers so that the reactive media minimize the constraints on groundwater flow;
- Environmental compatibility—Reactive media should be nontoxic and should not result in adverse chemical reactions or by-products when reacting with nitrate;
- Microbial carrier—Reactive media should have a large specific surface in order to serve as a solid carrier for biofilm formation.

On a basis of the above considerations, previous researchers conducted a large number of studies on reactive media. Wheat straw as a reactive medium could not persist for a long time because it lost 37.7 % of its mass during the 140-day operation (Saliling et al. 2007). Iron chips (30×200 mesh) could removed greater than 80 % of the influent NO_3-N, but iron and calcium precipitates led to cementation and reduction in permeability, iron corrosion and reactive surface (Phillips et al. 2000; Westerhoff and James 2003).

1.3.2 Field Application Cases

1.3.2.1 Biological Denitrification Permeable Reactive Barriers

BD PRBs have been successfully applied for remediation of nitrate contaminated groundwater. A BD PRB (35 m long, 1.5 m wide, 1.5 m deep) amended with sawdust (40 m^3) ran for 10 years in Cambridge, North Island, New Zealand (Schipper and Vojvodić-Vuković 1998, 2000, 2001; Schipper et al. 2004, 2005). Its denitrification rates (0.6–18.1 ng N/cm^3 h) were high enough to achieve nitrate removal (0.8–12.8 ng N/cm^3 h). Nevertheless most of groundwater flowed under rather than through the PRB. The reason(s) for this phenomenon are not well understood and are under investigation. Another BD PRB (1.2 m long, 0.6 m wide, 1.5 m deep) with coarse hardwood sawdust ran for 15 years in Long Point, Ontario, Canada (Robertson and Cherry 1995; Robertson et al. 2000). Over the 15 years of operation, the sawdust provided relatively complete removal of influent NO$_3$-N concentrations of up to 22.6 mg/L (Robertson et al. 2008). Denitrification consumed <1 % of the initial carbon mass annually (Robertson et al. 2000). Good denitrification capacity and ideal decadal longevity enhance the attractiveness of wood media for BD PRBs where very long-term, maintenance-free operation is required (Robertson et al. 2008).

1.3.2.2 Zero-valent Iron Permeable Reactive Barriers

ZVI PRBs have been successfully applied to remediate nitrate-contaminated groundwater as well. A ZVI PRB ($\sim$69 m in length, 0.9 m in wide, $\sim$9 m in depth) was installed to treat nitrate and other contaminants (e.g. Uranium) at the Oak Ridge Y-12 Plant site in Canada in late November 1997 (Gu et al. 2002). Denitrifying bacteria may have greatly increased the rate and extent of nitrate removal in the reducing zone of the barrier.

1.4 Conclusions

1. Groundwater is seriously contaminated by nitrate all over the world and its levels in aquifers are increasing.
2. Nitrate contamination is caused by nitrogenous fertilizers, livestock manures, agricultural irrigation, septic tanks, cesspools, pit latrines, contaminated land, river-aquifer interaction and atmospheric nitrogen deposition.
3. Nitrate is closely related to methemoglobinemia (referred to as blue-baby syndrome), hypertension, cancers, malformation, mutation, spontaneous abortion, and even death.
4. Without nitrate degradation, nitrate adsorbents (natural, synthetic and modified) will become saturated and lose the removal capacity.

5. A large number of liquid (such as VFAs and methanol), solid (such as cotton and sawdust) and gas (such as methane) organic carbons in HD have been evaluated, but cellulose-based substrates have great potential application prospects in *in situ* groundwater remediation.

6. The cell yield of approximately 0.24 g cells/g NO_3-N for Hydrogenotrophic denitrification is considerably lower than the 0.6–0.9 g cells/g NO_3-N for HD, but the addition of hydrogen to groundwater is not straightforward due to its low solubility. Fortunately, hydrogenotrophic denitrification can be sustained by various ZVI species.

7. Sulfur autotrophic denitrification has been studied for groundwater treatment, but low solubility of reduced sulfurs, sulfate production and biomass yield limit its applicability.

8. ZVI-based CR of nitrate is receiving considerable attention, but its main disadvantage is the release of ammonium as a major undesirable nitrogen product under acidic conditions.

9. More recently, a novel HAD approach was proposed in which the synergistic effects of HD and AD were encouraging. Nevertheless, further studies are needed to develop this approach.

10. PA, cellulose-based HD, ZVI-based CR and AD, and their combined approaches can be applied by means of PRB. So far, BD PRBs and ZVI PRBs have been successfully applied.

References

Ahn SC, Oh SY, Cha DK (2008) Enhanced reduction of nitrate by zero-valent iron at elevated temperatures. J Hazard Mater 156:17–22

Alabdula'aly AI, Al-Rehaili AM, Al-Zarah AI, Khan MA (2010) Assessment of nitrate concentration in groundwater in Saudi Arabia. Environ Monit Assess 161:1–9

Alowitz MJ, Scherer MM (2002) Kinetics of nitrate, nitrite, and Cr(VI) reduction by iron metal. Environ Sci Technol 36:299–306

Angelopoulos K, Spiliopoulos IC, Mandoulaki A, Theodorakopoulou A, Kouvelas A (2009) Groundwater nitrate pollution in northern part of Achaia Prefecture. Desalination 248:852–858

Aslan S, Turkman A (2003) Biological denitrification of drinking water using various natural organic solid substrates. Wat Sci Technol 48:489–495

Beneš V, Pěkný V, Skořepa J, Vrba J (1989) Impact of diffuse nitrate pollution sources on groundwater quality-some examples from Czechoslovakia. Environ Health Perspect 83:5–24

Bijay-Singh Yadvinder-Singh, Sekhon GS (1995) Fertilizer-N use efficiency and nitrate pollution of groundwater in developing countries. J Contam Hydrol 20:167–184

Blowes DW, Robertson WD, Ptacek CJ, Merkley C (1994) Removal of agricultural nitrate from tile-drainage effluent water using in-line bioreactors. J Contam Hydrol 15:207–221

Boley A, Müller W-R, Haider G (2000) Biodegradable polymers as solid substrate and biofilm carrier for denitrification in recirculated aquaculture systems. Aquacult Eng 22:75–85

Boumans L, Fraters D, van Drecht G (2004) Nitrate leaching by atmospheric N deposition to upper groundwater in the sandy regions of the Netherlands in 1990. Environ Monit Assess 93:1–15

Boumediene M, Achour D (2004) Denitrification of the underground waters by specific resin exchange of ion. Desalination 168:187–194

Bovolo CI, Parkin G, Sophocleous M (2009) Groundwater resources, climate and vulnerability. Environl Res Lett 4(3):1–12

Brennan RA, Sanford RA, Werth CJ (2006a) Biodegradation of tetrachloroethene by chitin fermentation products in a continuous flow column system. J Environ Eng-ASCE 132:664–673

Brennan RA, Sanford RA, Werth CJ (2006b) Chitin and corncobs as electron donor sources for the reductive dechlorination of tetrachloroethene. Water Res 40:2125–2134

Campos JL, Carvalho S, Portela R, Mosquera-Corral A, Méndez R (2008) Kinetics of denitrification using sulphur compounds: effects of S/N ratio, endogenous and exogenous compounds. Bioresour Technol 99:1293–1299

Chabani M, Bensmaili A (2005) Kinetic modelling of the retention of nitrates by Amberlite IRA 410. Desalination 185:509–515

Chang C, Tseng S, Huang H (1999) Hydrogenotrophic denitrification with immobilized *Alcaligenes eutrophus* for drinking water treatment. Bioresour Technol 69:53–58

Chen JY, Taniguchi M, Liu GQ, Miyaoka K, Onodera S-i, Tokunaga T, Fukushima Y (2007) Nitrate pollution of groundwater in the Yellow River delta, China. Hydrogeol J 15:1605–1614

Cheng IF, Muftikian R, Fernando Q, Korte N (1997) Reduction of nitrate to ammonia by zero-valent iron. Chemosphere 35:2689–2695

Choe S, Chang YY, Hwang KY, Khim J (2000) Kinetics of reductive denitrification by nanoscale zero-valent iron. Chemosphere 41:1307–1311

Choe SH, Ljestrand HM, Khim J (2004) Nitrate reduction by zero-valent iron under different pH regimes. Appl Geochem 19:335–342

Costa JL, Massone H, Martínez D, Suero EE, Vidal CM, Bedmar F (2002) Nitrate contamination of a rural aquifer and accumulation in the unsaturated zone. Agr Water Manag 57:33–47

Daniels L, Belay N, Rajagopal BS, Weimer PJ (1987) Bacterial methanogenesis and growth from CO_2 with elemental iron as the sole source of electrons. Science 237:509–511

Della Rocca C, Belgiorno V, Meric S (2005) Cotton-supported heterotrophic denitrification of nitrate-rich drinking water with a sand filtration post-treatment. Water SA 31:229–236

Della Rocca C, Belgiorno V, Meric S (2006) An heterotrophic/autotrophic denitrification (HAD) approach for nitrate removal from drinking water. Process Biochem 41:1022–1028

Della Rocca C, Belgiorno V, Meriç S (2007a) Overview of in situ applicable nitrate removal processes. Desalination 204:46–62

Della Rocca C, Belgiorno V, Meric S (2007b) Heterotrophic/autotrophic denitrification (HAD) of drinking water: prospective use for permeable reactive barrier. Desalination 210:194–204

Demiral H, Gündüzoğlu G (2010) Removal of nitrate from aqueous solutions by activated carbon prepared from sugar beet bagasse. Bioresour Technol 101:1675–1680

Deng DJ (2000) Progress of gastric cancer etiology: N-nitrosamides in the 1990s. World J Gastroentero 6:613–618

Devlin JF, Eedy R, Butler BJ (2000) The effects of electron donor and granular iron on nitrate transformation rates in sediments from a municipal water supply aquifer. J Contam Hydrol 46:81–97

Dong J, Zhao Y, Zhang W, Hong M (2009) Laboratory study on sequenced permeable reactive barrier remediation for landfill leachate-contaminated groundwater. J Hazard Mater 161:224–230

Eisentraeger A, Klag P, Vansbotter B, Heymann E, Dott W (2001) Denitrification of groundwater with methane as sole hydrogen donor. Water Res 35:2261–2267

Elefsiniotis P, Wareham DG (2007) Utilization patterns of volatile fatty acids in the denitrification reaction. Enzyme Microb Technol 41:92–97

Elmidaoui A, Elhannouni F, Taky M, Chay L, Elabbassi H, Hafsi M, Largeteau D (2002) Optimization of nitrate removal operation from ground water by electrodialysis. Sep Purif Technol 29:235–244

Ergas SJ, Reuss AF (2001) Hydrogenotrophic denitrification of drinking water using a hollow fibre membrane bioreactor. J Water Supply: Res Technol-AQUA 50:161–171

Fan AM, Steinberg VE (1996) Health implications of nitrate and nitrite in drinking water: an update on methemoglobinemia occurrence and reproductive and developmental toxicity. Regul Toxicol Pharm 23:35–43

Fernández-Nava Y, Marañón E, Soons J, Castrillón L (2010) Denitrification of high nitrate concentration wastewater using alternative carbon sources. J Hazard Mater 173:682–688

Gao W, Guan N, Chen J, Guan X, Jin R, Zeng H, Liu Z, Zhang F (2003) Titania supported Pd-Cu bimetallic catalyst for the reduction of nitrate in drinking water. Appl Catal B: Environ 46(2):341–351

Gavaskar AR (1999) Design and construction techniques for permeable reactive barriers. J Hazard Mater 68:41–71

Ghafari S, Hasan M, Aroua MK (2008) Bio-electrochemical removal of nitrate from water and wastewater—a review. Bioresour Technol 99:3965–3974

Gómez MA, González-López J, Hontoria-García E (2000) Influence of carbon source on nitrate removal of contaminated groundwater in a denitrifying submerged filter. J Hazard Mater 80:69–80

Greenan CM, Moorman TB, Kaspar TC, Parkin TB, Jaynes DB (2006) Comparing carbon substrates for denitrification of subsurface drainage water. J Environ Qual 35:824–829

Gu BH, Watson DB, Wu LY, Phillips DH, White DC, Zhou J (2002) Microbiological characteristics in a zero-valent iron reactive barrier. Environ Monit Assess 77:293–309

Guan H, Bestland E, Zhu C, Zhu H, Albertsdottir D, Hutson J, Simmons CT, Ginic-Markovic M, Tao X, Ellis AV (2010) Variation in performance of surfactant loading and resulting nitrate removal among four selected natural zeolites. J Hazard Mater 183(1–3):616–621

Haugen KS, Semmens MJ, Novak PJ (2002) A novel in situ technology for the treatment of nitrate contaminated groundwater. Water Res 36:3497–3506

Hu K, Huang Y, Li H, Li B, Chen D, White RE (2005) Spatial variability of shallow groundwater level, electrical conductivity and nitrate concentration, and risk assessment of nitrate contamination in North China plain. Environ Int 31:896–903

Huang YH, Zhang TC (2004) Effects of low pH on nitrate reduction by iron powder. Water Res 38:2631–2642

Huang CP, Wang HW, Chiu PC (1998) Nitrate reduction by metallic iron. Water Res 32:2257–2264

Hudak PF (2000) Regional trends in nitrate content of Texas groundwater. J Hydrol 228:37–47

Hunter WJ (2001) Use of vegetable oil in a pilot-scale denitrifying barrier. J Contam Hydrol 53:119–131

ITRC (2005) Permeable reactive barriers: lessons learned/new directions. Technical/Regulatory Guidelines, Washington, DC

Kaçaroğlu F, Günay G (1997) Groundwater nitrate pollution in an alluvium aquifer, Eskisehir urban area and its vicinity, Turkey. Environ Geol 31:178–184

Karanasios KA, Vasiliadou IA, Pavlou S, Vayenasa DV (2010) Hydrogenotrophic denitrification of potable water: a review. J Hazard Mater 180:20–37

Khan IA, Spalding RF (2004) Enhanced in situ denitrification for a municipal well. Water Res 38:3382–3388

Kielemoes J, De Boever P, Verstraete W (2000) Influence of denitrification on the corrosion of iron and stainless steel powder. Environ Sci Technol 34:663–671

Kim YS, Nakano K, Lee TJ, Kanchanatawee S, Matsumura M (2002) On-site nitrate removal of groundwater by an immobilized psychrophilic denitrifier using soluble starch as a carbon source. J Biosci Bioeng 93:303–308

Kim H, Seagren EA, Davis AP (2003) Engineered bioretention for removal of nitrate from stormwater runoff. Water Environ Res 75:355–367

Knowles R (2005) Denitrifiers associated with methanotrophs and their potential impact on the nitrogen cycle. Ecol Eng 24:441–446

Kumazawa K (2002) Nitrogen fertilization and nitrate pollution in groundwater in Japan: present status and measures for sustainable agriculture. Nutr Cycl Agroecosys 63:129–137

Lee K, Rittmann BE (2002) Applying a novel autohydrogenotrophic hollow-fiber membrane biofilm reactor for denitrification of drinking water. Water Res 36:2040–2052

Lee JW, Lee KH, Park KY, Maeng SK (2010) Hydrogenotrophic denitrification in a packed bed reactor: effects of hydrogen-to-water flow rate ratio. Bioresour Technol 101:3940–3946

Liao C-H, Kang S-F, Hsu Y-W (2003) Zero-valent iron reduction of nitrate in the presence of ultraviolet light, organic matter and hydrogen peroxide. Water Res 37:4109–4118

Liu H, Jiang W, Wan D, Qu J (2009) Study of a combined heterotrophic and sulfur autotrophic denitrification technology for removal of nitrate in water. J Hazard Mater 169:23–28

Luk GK, Au-Yeung WC (2002) Experimental investigation on the chemical reduction of nitrate from groundwater. Adv Environ Res 6:441–453

Matějů V, Čižinská S, Krejěí J, Janoch T (1992) Biological water denitrification—a review. Enzyme Microb Technol 14:170–183

Menkouchi Sahli MA, Annouar S, Mountadar M, Soufiane A, Elmidaouia A (2008) Nitrate removal of brackish underground water by chemical adsorption and by electrodialysis. Desalination 227:327–333

Mizuta K, Matsumoto T, Hatate Y, Nishihara K, Nakanishi T (2004) Removal of nitrate-nitrogen from drinking water using bamboo powder charcoal. Bioresour Technol 95:255–257

Modin O, Fukushi K, Yamamoto K (2007) Denitrification with methane as external carbon source. Water Res 41:2726–2738

Moon HS, Chang SW, Nam K, Choe J, Kim JY (2006) Effect of reactive media composition and co-contaminants on sulfur-based autotrophic denitrification. Environ Pollut 144:802–807

Moon HS, Shin DY, Nam K, Kim JY (2008) A long-term performance test on an autotrophic denitrification column for application as a permeable reactive barrier. Chemosphere 73:723–728

Moreno B, Gómez MA, González-López J, Hontoria E (2005) Inoculation of a submerged filter for biological denitrification of nitrate polluted groundwater: a comparative study. J Hazard Mater 117:141–147

Nolan BT (2001) Relating nitrogen sources and aquifer susceptibility to nitrate in shallow ground waters of the United States. Ground Water 39:290–299

Nolan BT, Hitt KJ (2006) Vulnerability of shallow groundwater and drinking-water wells to nitrate in the United States. Environ Sci Technol 40:7834–7840

Öztürk N, Bektas TE (2004) Nitrate removal from aqueous solution by adsorption onto various materials. J Hazard Mater 112:155–162

Peyton BM (1996) Improved biomass distribution using pulsed injections of electron donor and acceptor. Water Res 30:756–758

Phillips DH, Gu B, Watson DB, Lee SY (2000) Performance evaluation of a zerovalent iron reactive barrier: mineralogical characteristics. Environ Sci Technol 34:4169–4176

Prüsse U, Vorlop K-D (2001) Supported bimetallic palladium catalysts for water-phase nitrate reduction. J Mol Catal A: Chem 173:313–328

Raghoebarsing AA, Pol A, van de Pas-Schoonen KT, Smolders AJP, Ettwig KF, Rijpstra WIC, Schouten S, Sinninghe Damsté JS, Op den Camp HJM, Jetten MSM, Strous M (2006) A microbial consortium couples anaerobic methane oxidation to denitrification. Nature 440:918–921

Rajapakse JP, Scutt JE (1999) Denitrification with natural gas and various new growth media. Wat Res 33(18):3723–3734

Rao NS (2006) Nitrate pollution and its distribution in the groundwater of Srikakulam district, Andhra Pradesh, India. Environ Geol 51:631–645

Rezaee A, Godini H, Jorfi S (2010) Nitrate removal from aqueous solution using $MgCl_2$ impregnated activated carbon. Environ Eng Manag J 9:449–452

Rivett MO, Bussb SR, Morgan P, Smith JWN, Bemment CD (2008) Nitrate attenuation in groundwater: a review of biogeochemical controlling processes. Water Res 42:4215–4232

Robertson WD, Cherry JA (1995) In situ denitrification of septic-system nitrate using reactive porous media barriers: field trials. Ground Water 33(1):99–111

Robertson WD, Blowes DW, Ptacek CJ, Cherry JA (2000) Long-term performance of in situ reactive barriers for nitrate remediation. Ground Water 38:689–695

Robertson WD, Vogan JL, Lombardo PS (2008) Nitrate removal rates in a 15-year-old permeable reactive barrier treating septic system nitrate. Ground Water Monit Remediat 28:65–72

Robinson-Lora MA, Brennan RA (2009) The use of crab-shell chitin for biological denitrification: batch and column tests. Bioresour Technol 100:534–541

Rodríguez-Maroto JM, Garcia-Herruzo F, Garcia-Rubio A, Sampaio LA (2009) Kinetics of the chemical reduction of nitrate by zero-valent iron. Chemosphere 74:804–809

Ruangchainikom C, Liao CH, Anotai J, Lee M-T (2006) Characteristics of nitrate reduction by zero-valent iron powder in the recirculated and CO_2-bubbled system. Water Res 40:195–204

Ruckart PZ, Henderson AK, Black ML, Flanders WD (2008) Are nitrate levels in groundwater stable over time? J Expo Sci Environ Epidemiol 18:129–133

Salameh E, Alawi M, Batarseh M, Jiries A (2002) Determination of trihalomethanes and the ionic composition of groundwater at Amman City, Jordan. Hydrogeol J 10:332–339

Saliling WJB, Westerman PW, Losordo TM (2007) Wood chips and wheat straw as alternative biofilter media for denitrification reactors treating aquaculture and other wastewaters with high nitrate concentrations. Aquacult Eng 37:222–233

Salvestrin H, Hagare P (2009) Removal of nitrates from groundwater in remote indigenous settings in arid Central Australia. Desalin Water Treat 11:151–156

Sato Y, Murayama K, Nakai T, Takahashi N (1995) Nitric acid adsorption by a phosphonic acid ester type adsorbent. Water Res 29:1267–1271

Schipper LA, Vojvodić-Vuković M (1998) Nitrate removal from groundwater using a denitrification wall amended with sawdust: field trial. J Environ Qual 27:664–668

Schipper LA, Vojvodić-Vuković M (2000) Nitrate removal from groundwater and denitrification rates in a porous treatment wall amended with sawdust. Ecol Eng 14:269–278

Schipper LA, Vojvodić-Vuković M (2001) Five years of nitrate removal, denitrification and carbon dynamics in a denitrification wall. Water Res 35:3473–3477

Schipper LA, Barkle GF, Hadfield JC, Vojvodić-Vuković M, Burgess CP (2004) Hydraulic constraints on the performance of a groundwater denitrification wall for nitrate removal from shallow groundwater. J Contam Hydrol 69:263–279

Schipper LA, Barkle GF, Vojvodić-Vuković M (2005) Maximum rates of nitrate removal in a denitrification wall. J Environ Qual 34:1270–1276

Shin KH, Cha DK (2008) Microbial reduction of nitrate in the presence of nanoscale zero-valent iron. Chemosphere 72:257–262

Siantar DP, Schreier CG, Chou CS, Reinhard M (1996) Treatment of 1,2-dibromo-3-chloropropane and nitrate-contaminated water with zero-valent iron or hydrogen/palladium catalysts. Water Res 30:2315–2322

Sierra-Alvarez R, Beristain-Cardoso R, Salazar M, Gómez J, Razo-Flores E, Field JA (2007) Chemolithotrophic denitrification with elemental sulfur for groundwater treatment. Water Res 41:1253–1262

Smith RL, Ceazan ML, Brooks MH (1994) Autotrophic, hydrogen-oxidizing, denitrifying bacteria in groundwater, potential agents for bioremediation of nitrate contamination. Appl Environ Microbiol 60:1949–1955

Smith RL, Miller DN, Brooks MH, Widdowson MA, Killingstad MW (2001) In situ stimulation of groundwater denitrification with formate to remediate nitrate contamination. Environ Sci Technol 35:196–203

Soares MIM (2000) Biological denitrification of groundwater. Water Air Soil Pollut 123:183–193

Soares MIM, Abeliovich A (1998) Wheat straw as substrate for water denitrification. Water Res 32:3790–3794

Spalding RF, Exner ME (1993) Occurrence of nitrate in groundwater—a review. J Environ Qual 22:392–402

Starr RC, Gillham RW (1993) Denitrification and organic carbon availability in two aquifers. Ground Water 31:934–947

Steindorf K, Schlehofer B, Becher H, Hornig G, Wahrendorf J (1994) Nitrate in drinking-water. A case-control study on primary brain tumors with an embedded drinking water survey in Germany. Int J Epidemiol 23:451–457

Su C, Puls RW (2004) Nitrate reduction by zerovalent iron: effects of formate, oxalate, citrate, chloride, sulfate, borate, and phosphate. Environ Sci Technol 38:2715–2720

Su C, Puls RW (2007) Removal of added nitrate in the single, binary, and ternary systems of cotton burr compost, zerovalent iron, and sediment: implications for groundwater nitrate remediation using permeable reactive barriers. Chemosphere 67:1653–1662

Thalasso F, Vallecillo A, García-Encina P, Fernández-Polanco F (1997) The use of methane as a sole carbon source for wastewater denitrification. Water Res 31:55–60

Thiruvenkatachari R, Vigneswaran S, Naidu R (2008) Permeable reactive barrier for groundwater remediation. J Ind Eng Chem 14:145–156

Till BA, Weathers LJ, Alvarez PJJ (1998) Fe(0)-supported autotrophic denitrification. Environ Sci Technol 32:634–639

van Rijn J, Tal Y, Schreier HJ (2006) Denitrification in recirculating systems: theory and applications. Aquacult Eng 34:364–376

Volokita M, Belkin S, Abeliovich A, Soares MIM (1996a) Biological denitrification of drinking water using newspaper. Water Res 30:965–971

Volokita M, Abeliovich A, Soares MIM (1996b) Denitrification of groundwater using cotton as energy source. Wat Sci Technol 34(1–2):379–385

Wakida FT, Lerner DN (2005) Non-agricultural sources of groundwater nitrate: a review and case study. Water Res 39:3–16

Wang Q, Feng C, Zhao Y, Hao C (2009) Denitrification of nitrate contaminated groundwater with a fiber-based biofilm reactor. Bioresour Technol 100:2223–2227

Ward MH, deKok TM, Levallois P, Brender J, Gulis G, Nolan BT, VanDerslice J (2005) Workgroup report: drinking-water nitrate and health-recent findings and research needs. Environ Health Perspect 113:1607–1614

Westerhoff P, James J (2003) Nitrate removal in zero-valent iron packed columns. Water Res 37:1818–1830

Weyer PJ, Cerhan JR, Kross BC, Hallberg GR, Kantamneni J, Breuer G, Jones MP, Zheng W, Lynch CF (2001) Municipal drinking water nitrate level and cancer risk in older women: the Iowa women's health study. Epidemiology 12:327–338

WHO (2008) Guidelines for drinking water quality (3rd ed.)

Xi Y, Mallavarapu M, Naidu R (2010) Preparation, characterization of surfactants modified clay minerals and nitrate adsorption. Appl Clay Sci 48:92–96

Yang GCC, Lee HL (2005) Chemical reduction of nitrate by nanosized iron: kinetics and pathways. Water Res 39:884–894

Zhang TC, Lampe DG (1999) Sulfur: limestone autotrophic denitrification processes for treatment of nitrate-contaminated water: batch experiments. Water Res 33:599–608

Zhang WL, Tian ZX, Zhang N, Li XQ (1996) Nitrate pollution of groundwater in northern China. Agr Ecosyst Environ 59:223–231

Zhu ZL, Chen DL (2002) Nitrogen fertilizer use in China-contributions to food production, impacts on the environment and best management strategies. Nutr Cycl Agroecosys 63:117–127

Chapter 2
Heterotrophic-Autotrophic Denitrification

Abstract A novel heterotrophic-autotrophic denitrification (HAD) approach supported by granulated spongy iron, pine bark and mixed bacteria was proposed for remediation of nitrate contaminated groundwater in an aerobic environment. The HAD involves biological deoxygenation, chemical reduction (CR) of nitrate and dissolved oxygen (DO), heterotrophic denitrification (HD) and autotrophic denitrification (AD). The experimental results showed 0.121 d, 0.142 d and 1.905 d were needed to completely remove DO by HAD, spongy iron and mixed bacteria respectively. Spongy iron played a dominant role in deoxygenation in the HAD. After 16 days, NO_3-N removal was approximately 100, 6.2, 83.1, 4.5 % by HAD, CR, HD, AD, respectively. CR, HD and AD all contributed to the overall removal of NO_3-N, but HD was the most important denitrification mechanism. There existed symbiotic, synergistic and promotive effects of CR, HD and AD within the HAD. The different environmental parameters (e.g. water temperature) showed different effects on HAD. HAD was capable of providing steady denitrification rate (1.233–1.397 mg/L/d) for 3.5 months. Pine bark could provide sufficient organic carbon, spongy iron could steadily remove DO, and microbial activity maintained relatively constant. HAD denitrification was zero order with a reaction rate constant (K) of 1.3220 mg/L/d.

Keywords Groundwater remediation · Aerobic environment · Heterotrophic-autotrophic denitrification (HAD) · Chemical reduction (CR) · Heterotrophic denitrification (HD) · Autotrophic denitrification (AD) · Granulated spongy iron · Pine bark

Abbreviations

ACS	American Chemical Society
AD	Autotrophic denitrification
BD	Biological denitrification
BET	Brunauer-Emmett-Teller
CR	Chemical reduction
DO	Dissolved oxygen

F. Liu et al., *Study on Heterotrophic-Autotrophic Denitrification Permeable Reactive Barriers (HAD PRBs) for In Situ Groundwater Remediation*, SpringerBriefs in Water Science and Technology, DOI: 10.1007/978-3-642-38154-6_2, © The Author(s) 2014

DNRA   Dissimilatory nitrate reduction to ammonium
HAD   Heterotrophic-autotrophic denitrification
HD   Heterotrophic denitrification
HRT   Hydraulic retention time
MDR   Maximum denitrification rate
MNPR   Maximum NO_3-N percent removal
PRB   Permeable reactive barrier
RO   Reverse osmosis
TOC   Total organic carbon
ZVI   Zero-valent iron

2.1 Introduction

Nitrate contamination of groundwater has become an environmental and health issue in developed and developing countries (Della Rocca et al. 2007a). Nitrate contaminated groundwater often contains dissolved oxygen (DO) (≤ 7 mg/L) to form an aerobic environment (Gómez et al. 2002; Schnobrich et al. 2007).

Zero-valent iron (ZVI) in the forms of steel wool, nanoscale iron, granulated iron, cast iron, iron chips, iron powder, etc. has been worldwide reported to chemically or biotically remove nitrate from water (Cheng et al. 1997; Till et al. 1998; Huang et al. 1998; Westerhoff and James 2003; Choe et al. 2004; Yang and Lee 2005; Ahn et al. 2008; Rodríguez-Maroto et al. 2009). However, ZVI in the form of granulated spongy iron has been neglected.

Actually, the most economical, environmentally sound, promising and versatile approach being studied for nitrate denitrification is biological denitrification (BD). The majority of BD processes relies on heterotrophic denitrifers requiring an organic carbon source, however, groundwater has a low carbon content. As a result, an external organic carbon has to be supplied for bacterial growth. Recently, various studies have been conducted to evaluate the potential use of cellulose-rich solid organic carbon sources in heterotrophic denitrification (HD) processes, such as cotton, wood chips, wheat straw, newspaper, sawdust (Volokita et al. 1996a, b; Soares and Abeliovich 1998; Saliling et al. 2007; Kim et al. 2003; Robertson et al. 2008). Some of these carbon sources have been successfully contained in permeable reactive barrier (PRB) for in situ remediation or in reactors for *ex situ* remediation. However, valuable information on pine bark as a solid carbon substrate is lacking.

More recently, a heterotrophic-autotrophic denitrification (HAD) approach supported by steel wool and cotton was put forward by Della Rocca et al. (2006, 2007b). This HAD process involves chemical reduction (CR), HD and autotrophic denitrification (AD). Cotton could act as a source of organic carbon for HD; while carbon dioxide generated by HD could be employed as a source of inorganic carbon by autotrophic denitrifiers. Concomitantly, steel wool could reduce DO in water and produce cathodic hydrogen, which enhances both HD and AD. Soares et al. (2000) reported 1,200 kg of cotton was strongly compressed and water flowed through channels of lower resistance at a relatively high flow rate. As a consequence, the denitrification rate declined considerably. Therefore it's necessary to seek a "harder" matrix as a solid organic carbon source to both support biofilm development and optimize the hydrodynamic environment which favors HAD processes. Continuous column experiments were carried out to investigate the denitrification capacity, the effects of flow-rate, ZVI amount, inlet nitrate and phosphate on the HAD performance and the formation of by-products (such as iron and bacterial biomass) (Della Rocca et al. 2006, 2007b). Although high nitrate-nitrogen (NO_3-N) removal efficiencies were encouraging, questions arise regarding the deoxygenation capacities and pathways, the respective contributions of AD, HD, ZVI-based CR to the overall NO_3-N removal, the kinetics of NO_3-N denitrification, and the effects of environmental parameters (e.g. water temperature) on denitrification behavior in HAD processes.

In the above context, a new HAD approach supported by mixing granulated spongy iron, pine bark and mixed bacteria was proposed to remediate nitrate contaminated groundwater in an aerobic environment and further develop HAD processes. The objectives of this study were to: (1) explore the feasibility and efficiency of NO_3-N removal using the HAD process; (2) investigate the deoxygenation capacities and pathways of HAD; (3) determine the contributions of AD, HD, granulated spongy iron-based CR to the overall NO_3-N removal; (4) confirm the kinetics of NO_3-N denitrification; (5) evaluate the effects of mass of pine bark, water temperature, high-concentration nitrate, nitrite, ammonium and coexistent inorganic anions on the denitrification behavior in the HAD; and (6) provide information for future column studies and field in situ applications.

2.2 Materials and Methods

2.2.1 Materials and Chemicals

Granulated spongy iron (60.60 % of Fe^0; 0.425–1.000 mm in diameter) was obtained from Kaibiyuan Co., Beijing, China. Pine bark (0.15–20.00 mm in diameter) was obtained from a local nursery store in Adelaide, South Australia. Moist sub-surface soil (0.3 m depth from surface) was taken from a pristine and humic-acid-rich area on the campus of the Flinders University of South Australia. Milli-Q (Millipore) water was used to prepare reagent solutions. Reverse osmosis (RO) water was spiked with mineral salts media acting as synthetic groundwater. In addition to NO_3-N (10–100 mg/L), the mineral salts media contained the following (final concentration in mg/L): $NaHCO_3$ (482), K_2HPO_4 (17.4), $FeCl_3 \cdot 6H_2O$ (0.53), Na_2EDTA (7.4), $MgCl_2 \cdot 6H_2O$ (40.6), $MgSO_4 \cdot 7H_2O$ (49), $CaCl_2 \cdot 6H_2O$

(21.9), NaCl (58.5), and trace elements which was composed of $Na_2MoO_4 \cdot 2H_2O$ (0.504), $CoCl_2 \cdot 6H_2O$ (0.08), $ZnSO_4 \cdot 10H_2O$ (0.088), $MnCl_2 \cdot 4H_2O$ (0.72) (Huang et al. 2012). Unless otherwise indicated, all chemicals used were analytical or American Chemical Society (ACS) reagent grade as received.

2.2.2 Enrichment Culture Protocol to Establish a Denitrifying Bacterial Population

The soil was passed through 0.425 mm and 0.075 mm sieves and then stored in a cold room (3.2 ± 1.0 °C) before use (Huang et al. 2012). Initial enrichment culture was undertaken using a 1 L Schott bottle to which was added: (1) 5.00 g of soil; (2) 800 ml of RO water which was enriched with 22.6 mg NO_3-N/L and 3.0 mg K_2HPO_4-P/L; (3) 5.00 g of pine bark; (4) 5.00 g of granulated spongy iron. The N:P weight ratio was 22.6:3 and initial DO was 3.95 mg/L. Following initial enrichment the denitrifying population was maintained by successively subculturing. Upon depletion of NO_3-N a bacterial suspension was transferred into fresh mineral salts media containing spongy iron and pine bark without soil at a bacterial suspension: mineral salts media volume ratio of 1:9. The enrichment culture was conducted in water bath (15 °C).

2.2.3 Batch Experiments

Batch incubations were conducted in sterile, screw capped, 2 L Schott bottles, to which was added mineral salts media. Spongy iron was sterilized (160 °C for 4 h). All the glassware, pipette tips, storage containers and mineral salts media were steam autoclaved (121 °C for 15–17 min) prior to use (Huang et al. 2012). When BD, biological deoxygenation and environmental parameters were being assessed an inoculum with mixed bacteria from an enriched denitrifying bacterial subculture was used, and the inoculum accounted for 10 % of the total mineral salts media volume (Huang et al. 2012). The granulated spongy iron: mineral salts solution ratio of 5:800 (mg:ml) was maintained. The Schott bottles were equipped with membrane screw caps and covered with aluminium foil and statically incubated in the dark (Huang et al. 2012). Water samples were drawn with a syringe and filtered through a 0.45 μm glass fibre filter (Whatman GF/C, 47 mm in diameter). The filtrate was placed in a screw capped plastic tube and stored at 3.2 ± 1.0 °C until analysis for NO_3-N, nitrite-nitrogen (NO_2-N), ammonium-nitrogen (NH_4-N), pH and T. DO was monitored online in the incubation bottles. The experimental conditions of batch incubations are shown in Table 2.1.

Table 2.1 The design of batch incubations used in this study

Incubations	Granulated spongy iron (g)	Pine bark (g)	Inoculum (ml)	Initial NO$_3$-N (mg/L)	Initial DO (mg/L)	Mineral salts media (g)	Water temperature (°C)	Initial pH
Deoxygenation by HAD	12.50	12.50	200	0	4.2	2000	15	8.04
Deoxygenation by CR	12.50	0	0	0	4.7	2000	15	7.87
Deoxygenation by mixed bacteria	0	12.50	200	0	4.1	2000	15	8.04
NO$_3$-N removal by HAD	12.50	12.50	200	20.35	4.3	2000	15	7.83
NO$_3$-N removal by HAD (control)	0	12.50	0	21.96	4.3	2000	15	7.83
NO$_3$-N removal by CR	12.50	0	0	22.57	0	2000	15	8.32
NO$_3$-N removal by HD	0	12.50	200	21.64	0	2000	15	7.72
NO$_3$-N removal by AD + CR	12.50	0	200	23.46	4.3	2000	15	8.32
Effect of the mass of pine bark on HAD	12.50	2.50	200	21.09	4.3	2000	15	8.16
		7.50		21.53				
		12.50		22.00				
Effect of water temperature on HAD	12.50	12.50	200	21.32	4.5	2000	15	8.16
				21.47			27.5	
				22.06			33	
Effect of high-concentration nitrate on HAD	12.50	12.50	200	97.58	4.5	2000	15	8.16
				20.35				
Effect of nitrite on HAD	12.50	12.50	200	20.48	4.5	2000	15	8.16
Effect of ammonium on HAD	12.50	12.50	200	21.46	4.5	2000	15	8.16
Effect of coexistent inorganic cations on HAD	12.50	12.50	200	21.99	4.5	2000	15	8.16
				–				
				22.56				

(continued)

Table 2.1 (continued)

Incubations	Granulated spongy iron (g)	Pine bark (g)	Inoculum (ml)	Initial NO_3-N (mg/L)	Initial DO (mg/L)	Mineral salts media (g)	Water temperature (°C)	Initial pH
Longevity of pine bark	12.50	12.50	200	20.77	4.5	2000	15	8.16
Kinetics of NO_3-N denitrification by HAD	12.50	12.50	200	23.62 44.20 20.35 11.15	4.1	2000	15	8.08
Kinetics of NO_3-N denitrification by HAD (control)	0	0	0	20.35	4.1	2000	15	8.08

2.2.4 Analytical Methods

Inorganic nitrogen (NO_3-N, NO_2-N and NH_4-N) was analysed using a FOSS-Tecator FIAStar 5,000 flow injection analyser (Sweden) equipped with a FOSS 5027 Auto-Sampler, a nitrite/nitrate method cassette, a ammonium method cassette (including gas diffusion cell and gas diffusion membrane), a 40 μl sample loop, a 400 μl sample loop, a cadmium reduction column and interference filters (M = 590 nm and R = 720 nm; M = 540 nm and R = 720 nm) employing American Public Health Association Standard Methods (APHA et al. 1992). The lower detection limits for NO_3-N, NO_2-N and NH_4-N were 0.005, 0.005 and 0.010 mg/L respectively. Temperature, pH, and DO were measured using a pH meter (Hanna, Model 8417, Italy) and a digital DO meter (Hanna, Model 9143, Italy).

2.2.5 Data Processing Methods

Variation of NO_3-N (or NO_2-N or NH_4-N or DO) (Eq. 2.1)

$$\triangle C = C_t - C_0 \tag{2.1}$$

where, $\triangle C$ is variation of NO_3-N (or NO_2-N or NH_4-N or DO) concentration (mg/L); t is reaction time (d); C_t is NO_3-N (or NO_2-N or NH_4-N or DO) concentration at time t (mg/L); C_0 is initial NO_3-N (or NO_2-N or NH_4-N or DO) concentration at time 0 (mg/L)NO_3-N (or NO_2-N or NH_4-N or DO) percent removal (Eq. 2.2)

$$\sigma = (C_t - C_0)/C_t \times 100\% \tag{2.2}$$

where, σ is NO_3-N (or NO_2-N or NH_4-N or DO) percent removal (%) Variation of pH (Eq. 2.3)

$$\triangle pH = pH_t - pH_0 \tag{2.3}$$

where, $\triangle pH$ is variation of pH; pH_t is pH at time t; pH_0 is initial pH at time 0 NO_3-N (or NO_2-N) denitrification rate (Eq. 2.4)

$$\Phi = (C_t - C_0)/t \tag{2.4}$$

where, φ is NO_3-N (or NO_2-N) denitrification rate (mg/L/d).

2.3 Results and Discussion

2.3.1 Deoxygenation Capacity of the Heterotrophic-Autotrophic Denitrification Process

To apply HAD processes to nitrate contaminated groundwater, the initial DO in raw water must be eliminated to create an anaerobic environment for chemical nitrate reduction and biodenitrification. Dissolved oxygen is thermodynamically more reducible than dissolved nitrate by ZVI due to a higher standard cell-reaction potential of oxygen reduction than that of nitrate reduction (Su and Puls 2007). On the other hand, BD produces less energy yield than oxygen respiration, therefore, a bacterial cell growing in aerobic environments will choose to use oxygen as a terminal electron acceptor (Gómez et al. 2002). In addition to this competitive effect, oxygen negatively controls BD at two levels: reversible inhibition of the activities of denitrification enzymes and regulation of gene expression (Gómez et al. 2002).

Figure 2.1 shows the changes in DO percent removal in mineral salts media by HAD, granulated spongy iron only, and mixed bacteria only. Linear regression analysis was used to describe DO removal over reaction time and estimate the time when complete reduction of DO was achieved. Strong positive linear correlations between percent removal and reaction time in the HAD, granulated spongy iron only, and mixed bacteria only incubations were determined (Eqs. 2.5, 2.6, and 2.7; Fig. 2.1).

For HAD:

$$\sigma(\%) = 793.0137t(\mathrm{d}) + 4.2059 \left(R^2 = 0.9854, \ P < 0.001 \right) \qquad (2.5)$$

For granulated spongy iron:

$$\sigma(\%) = 702.3783t - 0.4572 \left(R^2 = 0.9951, \ P < 0.001 \right) \qquad (2.6)$$

Fig. 2.1 Removal of dissolved oxygen (DO) by heterotrophic-autotrophic denitrification (HAD), granulated spongy iron and mixed bacteria at the initial DO of 4.2, 4.7 and 4.1 mg/L, respectively

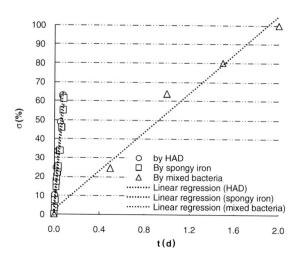

For mixed bacteria:

$$\sigma(\%) = 51.1922t + 2.4818 \left(R^2 = 0.9781, \ P < 0.001 \right) \tag{2.7}$$

Linear least square analysis demonstrated that 0.121 d, 0.142 d and 1.905 d were needed to completely remove DO by HAD, granulated spongy iron and mixed bacteria respectively. Clearly, HAD, granulated spongy iron, and mixed bacteria all had deoxygenation capacity, and the mixed bacteria contained aerobic heterotrophs. Therefore, granulated spongy iron and aerobic heterotrophs both contributed to DO removal in the HAD incubation. Granulated spongy iron consumes DO via CR (Eq. 2.8) (Della Rocca et al. 2005a). Aerobic heterotrophs employed total organic carbon (TOC) released by pine bark to deoxygenate via aerobic respiration (Eq. 2.9). Undoubtedly, HAD depended on granulated spongy iron via CR and aerobic heterotrophs via aerobic respiration to remove DO. Any incubation that had granulated spongy iron included reduced DO more rapidly than the mixed bacteria alone (Fig. 2.1). Consequently, granulated spongy iron played a dominant role in deoxygenation in the HAD (Huang et al. 2012).

$$2Fe^0 + O_2 + 2H_2O \rightarrow 2Fe^{2+} + 4OH^- \tag{2.8}$$

$$2C_6H_{10}O_2 + 15O_2 \rightarrow 12CO_2 + 10H_2O \tag{2.9}$$

Oxygen reduction is considered necessary prior to the onset of nitrate removal. The HAD process has shown the rapid deoxygenation capacity (Fig. 2.1). Therefore, this process facilitated an anaerobic environment which avoids the competitive and inhibitory effects of DO. This is a desirable attribute because most nitrate laden groundwater also contains oxygen (Huang et al. 2012).

2.3.2 Contributions of Chemical Reduction, Heterotrophic Denitrification and Autotrophic Denitrification to the Performance of the Heterotrophic-Autotrophic Denitrification Process

2.3.2.1 Performance of Heterotrophic-Autotrophic Denitrification

Figure 2.2 shows the changes in NO_3-N, NO_2-N, NH_4-N and pH in the HAD incubation.

As shown in Fig. 2.2a, NO_3-N decreased with increasing reaction time. 19.92 mg/L NO_3-N was removed during the 16 day period (Fig. 2.2a), suggesting that the HAD approach not only was effective at removing NO_3-N but also is worthy of further study. 0.27 mg/L NO_3-N was lost from the control incubation for NO_3-N in 16 days (Fig. 2.2a). This indicated that the effect of NO_3-N physical adsorption onto pine bark was negligible, i.e., NO_3-N was not removed by adsorption in the HAD incubation. NO_2-N tended to reach its maximum (0.15 mg/L) at day 10, and

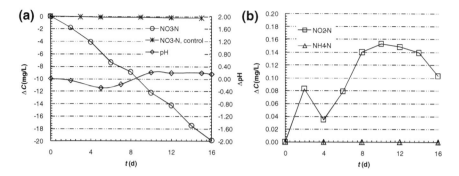

Fig. 2.2 Performance of heterotrophic-autotrophic denitrification (HAD) in the HAD incubation: **a** variations of NO₃-N and pH; **b** variations of NO₂-N and NH₄-N. The control incubation contained NO₃-N of 21.96 mg/L and pine bark of 12.50 g (sterilized, 160 °C for 4 h) without granulated spongy iron and mixed bacteria

subsequently decreased to 0.10 mg/L at day 16 (Fig. 2.2b). NO₂-N accounted for ≤0.77 % of the NO₃-N removed. No NH₄-N variations were found (Fig. 2.2b). Actually, NH₄-N concentrations were below the detection limit at all times. As a result, the HAD approach transformed nitrate into gaseous nitrous compounds (such as NO, N₂O) or nitrogen gas (N₂). The small variations in pH ranging from −0.30 to 0.20 were noted at the initial pH of 7.83 between days 0 and 16 (Fig. 2.2a). It is likely that the bicarbonate in the mineral salts media acting as a pH buffer controls the pH change, but no measurements were made to confirm this. As is well known, pH increase is probably detrimental to microbial metabolism and may limit nitrate and particularly nitrite BD. In combination with the nitrate and nitrite changes, it can be concluded that the solution pH never inhibited the HAD performance. For the case of groundwater remediation, it's possible that a denitrification PRB filled with spongy iron and pine bark could be installed in aquifers. Obviously, the HAD approach was potentially a feasible and effective approach for in situ groundwater remediation.

2.3.2.2 Performance of Chemical Reduction Denitrification

Figure 2.3 shows the changes in NO₃-N and pH in the CR incubation. Only 1.34 mg/L NO₃-N was removed during the 16 day period at the initial pH of 8.32 (Fig. 2.3), demonstrating that the capacity of nitrate CR by granulated spongy iron was limited at weakly alkaline pHs, which was consistent with the literature describing CR (Westerhoff and James 2003). The reason for the low reduction capacity was a limited supply of electrons which come from granulated spongy iron either directly or indirectly via the acidic corrosion products of Fe such as hydrogen. Part of Fe⁰ on the surface of granulated spongy iron was oxidized in the deoxygenation process, and then granulated spongy iron was covered with a passive oxide film consisting of an inner layer of Fe₃O₄ and an outer passive layer

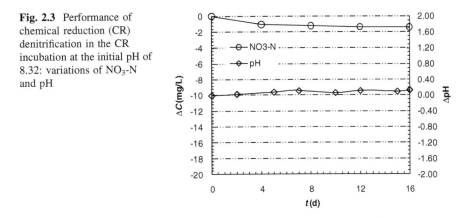

Fig. 2.3 Performance of chemical reduction (CR) denitrification in the CR incubation at the initial pH of 8.32: variations of NO_3-N and pH

Table 2.2 Performance of chemical reduction (CR) denitrification in the CR incubation at the initial pH of 8.32: variations of NO_2-N and NH_4-N

	0 d	4 d	8 d	12 d	16 d
ΔC (NO_2-N) (mg/L)	0.00	0.00	0.01	0.01	0.00
ΔC (NH_4-N) (mg/L)	0.00	0.00	0.01	0.02	0.00

of Fe_2O_3 (Tsai et al. 2009). This passive layer prevented the onset of electron transfer (Tsai et al. 2009). In weakly alkaline solutions, possible pathways of chemical NO_3-N reduction by Fe^0 are proposed in Eqs. 2.10, 2.11, and 2.12 (Siantar et al. 1996; Kielemoes et al. 2000). Table 2.2 illustrates the changes in NO_2-N and NH_4-N in the CR incubation. Almost no NO_2-N and NH_4-N variations were observed (Table 2.2). This implied that the NO_3-N removed was converted to gaseous nitrogen. It has been well demonstrated and generally accepted that regardless of Fe^0 types, rapid and complete nitrate CR mainly occurs at acidic pH $\leq$ 4 or 5, and neutral or alkaline conditions are not normally favorable for nitrate reduction (Huang et al. 1998; Huang and Zhang 2004; Yang and Lee 2005; Rodríguez-Maroto et al. 2009), but the main disadvantage of low pHs is the release of ammonium as a major undesirable nitrogen product.

$$5Fe^0 + 2NO_3^- + 6H_2O \rightarrow 5Fe^{2+} + N_2 + 12OH^- \tag{2.10}$$

$$10Fe^0 + 6NO_3^- + 3H_2O \rightarrow 5Fe_2O_3 + 3N_2 + 6OH^- \tag{2.11}$$

$$4Fe^0 + NO_3^- + 7H_2O \rightarrow 4Fe^{2+} + NH_4^+ + 10OH^- \tag{2.12}$$

$$Fe^0 + 2H_2O \rightarrow H_2 + Fe^{2+} + 2OH^- \tag{2.13}$$

Theoretically, the rise in pH can be achieved from Fe^0-based chemical nitrate reduction as well as anaerobic Fe^0 corrosion (Eqs. 2.10, 2.11, 2.12, and 2.13).

Accordingly, the minor rise in pH ($\triangle$pH $= 0 \sim 0.13$) (Fig. 2.3) could also indicate that the CR process did not progress significantly via single granulated spongy iron at weakly alkaline pHs.

2.3.2.3 Performance of Heterotrophic Denitrification

Figure 2.4 shows the changes in NO_3-N, NO_2-N, NH_4-N and pH in the HD incubation under an anaerobic condition. 17.97 mg/L NO_3-N was depleted during the 16 day period (Fig. 2.4a), demonstrating that the mixed bacteria contained heterotrophic denitrifiers. Heterotrophic denitrifiers could utilize TOC released by pine bark to reduce NO_3-N, for synthesis of new bacterial cells and maintenance of the existing cell mass (Eq. 2.14) (Ovez et al. 2006; Della Rocca et al. 2005b). NO_2-N concentration was less than 0.08 mg/L (Fig. 2.4b). Hence, no nitrite accumulation occurred. Nitrite accumulation during HD has previously been reported. NO_2-N concentration did not drop below 13 mg/L in a batch experimental study where pine shavings served as a solid carbon source. In other studies, nitrite accumulated when sucrose was used as a liquid carbon source (Piñar and Ramos 1998; Gómez et al. 2000). However, ethanol and methanol as liquid carbon sources did not cause the occurrence of nitrite accumulation (Gómez et al. 2000). It can be deduced that nitrite accumulation seems to depend on carbon source types. No NH_4-N variations were observed (Fig. 2.4b).

$$C_6H_{10}O_2 + 6NO_3^- + 6H^+ \rightarrow 6CO_2 + 8H_2O + 3N_2 \tag{2.14}$$

According to Eq. 2.14, pH should increase in HD as a result of the consumption of hydrogen ions ($H^+\downarrow$). But the results suggested just the opposite at the initial pH of 7.72 (Fig. 2.4a). The solution pH was between 7 and 8, indirectly indicating the heterotrophic denitrifying bacteria functioned effectively. It is widely believed that heterotrophic denitrifying microbial activity is at an optimal around neutrality (pH 6.0–8.0) (Beaubien et al. 1995; Elefsiniotis and Li 2006).

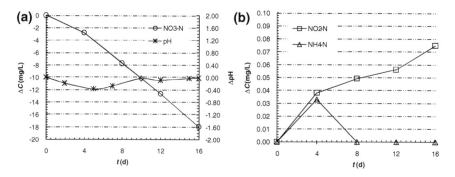

Fig. 2.4 Performance of heterotrophic denitrification (HD) in the HD incubation under an anaerobic condition: **a** variations of NO_3-N and pH; **b** NO_2-N and NH_4-N. The mineral salts media used was sparged with N_2 gas for 5 min at 100 kPa after sampling

2.3.2.4 Performance of Autotrophic Denitrification

Considering the explosion and safety concerns associated with hydrogen gas, H_2 was not employed to investigate the performance of AD. AD was identified by investigating the difference in performance between of AD coupled with CR (AD + CR) and CR only. Likewise, pH caused by AD at time t ($pH_{t,AD}$) was obtained by determination of the difference between AD + CR ($pH_{t,AD+CR}$) and CR ($pH_{t,CR}$). To preclude the probable interference of TOC and suspended solid from the mixed inoculum, the procedures were introduced as follows: first, filter the inoculum suspension through a 0.45 μm fiber filter membrane; second, filter the filtrate through a 0.2 μm filter membrane; third, add the 0.2 μm membrane into a AD + CR incubation without pine bark. Figure 2.5 shows the changes in NO_3-N, NO_2-N, NH_4-N and pH caused by AD and also shows their changes in the AD + CR incubation.0.87 mg/L NO_3-N was removed during the 16 day period (Fig. 2.5a), indicating that AD occurred but also the mixed bacteria contained autotrophic denitrifiers.

Under anaerobic environments, the corrosion of spongy iron by water formed cathodic hydrogen and ferrous iron (Eq. 2.13) (Choe et al. 2000; Su and Puls 2004). Hydrogenotrophic denitrifying bacteria utilized hydrogen and inorganic carbon (CO_2, HCO_3^-, etc.) to reduce NO_3-N (Haugen et al. 2002). The *beta-Proteobacteria*, including *Rhodocyclus*, *Hydrogenophaga*, and *beta- Proteobacteria HTCC 379*, probably played an important role in hydrogenotrophic denitrification (Zhang et al. 2009). Also, a group of lithoautotrophic bacteria might couple an AD process to the oxidation of Fe^{2+} with an inorganic carbon source (Straub and Buchholz-Cleven 1998; Weber et al. 2006). Additionally, the spongy iron used served as a solid-carrier for biofilm formation. Autotrophic denitrifying growth on Fe^0 has been previously demonstrated (Till et al. 1998). Little NO_2-N ($\leq$0.04 mg/L) and NH_4-N ($\leq$0.06 mg/L) were generated via AD (Fig. 2.5b).

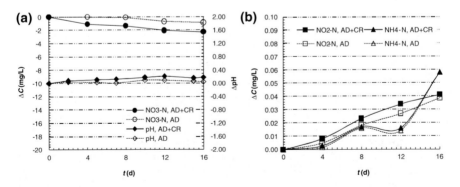

Fig. 2.5 Performance of autotrophic denitrification (AD) and autotrophic denitrification coupled with chemical reduction (AD + CR): **a** variations of NO_3-N and pH; **b** NO_2-N and NH_4-N. For NO_3-N (or NO_2-N or NH_4-N), $C_{t,AD}$ (mg/L) = $C_{t,AD+CR}$ (mg/L)-$C_{t,CR}$ (mg/L) at time t; For pH, $pH_{t,AD} = pH_{t,AD+CR}$-$pH_{t,CR}$ at time t

While pH values ranging from 7.0 to 8.0 are optimal for HD, previous researchers have observed that hydrogenotrophic denitrification did not appear to be inhibited by pH values up to approximately 8.6. Both abiotic and biotic NO_3-N reduction result in the elevation of pH in water due to either the production of hydroxide ions ($OH^-\uparrow$) or the consumption of hydrogen ions ($H^+\downarrow$) in Fe^0-supported hydrogenotrophic denitrification (Eqs. 2.13, 2.15, 2.16, and 2.17).

$$2NO_3^- + 5H_2 \rightarrow N_2 + 4H_2O + 2OH^- \tag{2.15}$$

$$2NO_3^- + 2H^+ + 5H_2 \rightarrow N_2 + 4H_2O \tag{2.16}$$

$$H_2 + 0.35NO_3^- + 0.35H^+ + 0.052CO_2 \rightarrow 0.17N_2 + 1.1H_2O + 0.010C_5H_7O_2N \tag{2.17}$$

The minor rise in pH ($\triangle$pH 0 $\sim$ 0.21 for AD + CR; $\triangle$pH 0 $\sim$ 0.13 for CR) (Fig. 2.5a) at the initial pH of 8.32 therefore appeared to demonstrate that the AD + CR and CR only processes did not progress significantly at weakly alkaline pHs. It's well known that Fe^0 corrosion is slow at pH $\geq$ 7 and autotrophic denitrifiers grow slowly, causing the low capacity of the AD.

2.3.2.5 Contributions of Chemical Reduction, Heterotrophic Denitrification and Autotrophic Denitrification

Figure 2.6 shows the changes in NO_3-N percent removal in mineral salts media by granulated spongy iron-based CR only, HD only, AD + CR, HAD, AD and HD + AD + CR. Percent removal of NO_3-N for AD was defined by percent removal for AD + CR minus percent removal for HD. Percent removal of NO_3-N for HD + AD + CR was defined by percent removal for AD + CR plus percent removal for HD.

The percent removal of NO_3-N escalated with increasing reaction time in the CR, HD, AD + CR and HAD incubations (Fig. 2.6). After 16 days little NO_3-N was lost from the CR incubation (6.2 % removed) at the initial pH 8.32. After 16 days there was large loss of NO_3-N in the HD incubation (83.1 % removed). This value was greater than 6.2 % in the single CR incubation, indicating AD not only occurred but also reduced 4.5 % of the intial NO_3-N. After 16 days NO_3-N removal was approximately 100 % in the HAD incubation.

There is no doubt that CR, HD and AD all contributed to the overall removal of NO_3-N in the HAD incubation in 16 days, demonstrating their symbiotic, synergistic and promotive effects within the HAD. The NO_3-N losses were greater for HD than for CR + AD, confirming that HD was the most important NO_3-N denitrification mechanism in the HAD (Huang et al. 2012). It is interesting to note that the combined percent removal for AD + CR + HD was slightly less than that for HAD at any point in time. The largest difference (4.2 %) between them was attained at day 16. An explanation is that in the HAD incubation, carbon dioxide

Fig. 2.6 The percent removal of NO$_3$-N by chemical reduction (CR), heterotrophic denitrification (HD), autotrophic denitrification coupled with chemical reduction (AD + CR), heterotrophic-autotrophic denitrification (HAD) and the calculated percent removal of NO$_3$-N by autotrophic denitrification (AD) and *HD + AD + CR*

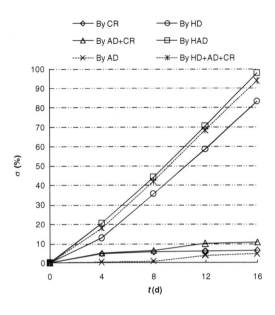

generated by aerobic heterotrophs and hetetrophic denitrifers was consumed by autotrophic denitrifiers as an additional inorganic carbon source to enhance the capacity of AD (Eqs. 2.9, and 2.14) (Della Rocca et al. 2006; Ovez et al. 2006). Concomitantly, the consumption of molecular hydrogen overcomed the decrease of available reactive surface area of Fe^0 (Choe et al. 2004).

2.3.3 Effects of Environmental Parameters on the Heterotrophic-Autotrophic Denitrification Process

2.3.3.1 Effect of the Mass of Pine Bark

As mentioned in Sect. 2.3.2.5, the majority of NO$_3$-N removal by the HAD relied on HD that necessarily requires an organic carbon source, so the mass of pine bark might affect the HAD performance. Three different amounts of pine bark (2.5, 7.5 and 12.5 g) were incubated in a constant mineral salts media (2.0 L). A comparison was made of denitrification rates among the three amounts in Fig. 2.7a. The changes in NO$_2$-N and NH$_4$-N over reaction time at the three weights of pine bark are shown in Fig. 2.7b.

In the first 4 days of incubation, the denitrification rates for the three amounts were almost the same. There was a minor effect of pine bark amount on the HAD denitrification during this period of time (Fig. 2.7a). This is attributed to the acclimation of the denitrifying bacteria to this substrate (Elefsiniotis and Li 2006).

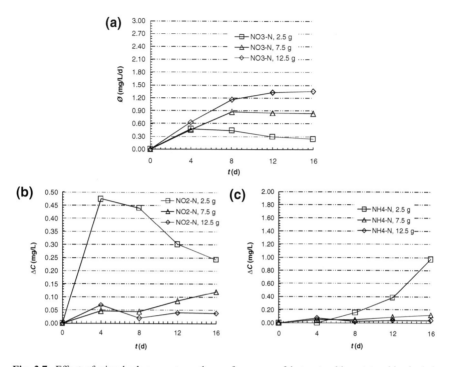

Fig. 2.7 Effect of pine bark amount on the performance of heterotrophic-autotrophic denitrification (HAD) over reaction time at 2.5, 7.5 and 12.5 g of pine bark in 2000 mL of mineral salts meida: **a** denitrification rates; **b** variations of NO_2-N; **c** variations of NH_4-N

In the last 4 days of incubation, the denitrication rate increased in the order of pine bark weight: 12.5 g > 7.5 g > 2.5 g (Fig. 2.7a). There was a noticeable effect of pine bark mass on the HAD performance from days 4 to 16. Larger amounts of pine bark could provide more organic carbon and electron donors for heterotrophic bacterial respiration and growth. Besides, increasing the pine bark mass could increase the surface area of pine bark available to the denitrifying population (Ovez et al. 2006). Overall, denitrification rates were closely and positively related to the mass of pine bark.

The most significant NO_2-N accumulation at day 4 at the pine bark of 2.5 g (Fig. 2.7b) resulted from a lack of pine bark supply. Limitation of organic carbon causes the competition between nitrite and nitrate reductase for electron donors, and induces delayed syntheses and inhibited activities of nitrite reductase relative to nitrate reductase (Hunter 2003). These leads to denitrification rates of NO_3-N being lower than those of NO_2-N (Betlach and Tiedje 1981), which gives rise to the incomplete reduction of NO_3-N.

Low dissolved NH_4-N concentrations of ≤ 0.12 mg/L were found at the pine bark of 7.5 and 12.5 g, but high concentrations of 0.38–0.97 mg/L, which accounted for 10.89–24.95 % of the NO_3-N transformed, were observed during the last 4 days of incubation at the pine bark of 2.5 g (Fig. 2.7c). Dissimilatory nitrate reduction to ammonium (DNRA) by microorganisms might be a mechanism of NH_4-N production (Kim et al. 2003). However, DNRA requires a carbon rich and high organic carbon to nitrate environment (Yin et al. 2002; Robertson et al. 2008). Therefore it is unlikely that DNRA was the major cause of NH_4-N production in the incubation with the smallest weight of pine bark; notwithstanding it only accounted for <1 % of the NO_3-N transformed in HD systems (Greenan et al. 2006). Elucidation of NH_4-N production in the HAD is not a primary focus of this study and the exact cause remains unknown.

2.3.3.2 Effect of Water Temperature

Water temperature has different effects on different BD processes. Volokita et al. (1996a) reported that denitrification rates at 14 °C were approximately half of the rates observed at 30 °C when cotton was used as a carbon source for HD. Volokita et al. (1996b) reported denitrification rates at 14 °C were approximately one third of the rates at 32 °C when newspaper was used as another carbon source. Robertson et al. (2000) reported that denitrification rates at 2–5 °C were 5 mg N/L/d, whereas the corresponding rates at 10–20 °C were 15–30 mg N/L/d, when coarse wood mulch acted as a carbon source (Robertson et al. 2000). Huang et al. (2012) reported denitrification rate at 27.5 °C was 1.36 times higher than at 15.0 °C for a HAD denitrification with methanol and spongy iron. It's not clear how water temperature influences the HAD process with pine bark and spongy iron from literature. The optimum temperature for BD is between 25 and 35 °C, and meanwhile denitrificaiton can take place in the range 2–50 °C due to the bacteria capacity to survive in extreme environmental conditions (Karanasios et al. 2010). In the study reported here, a cold temperature (15 °C) was chosen based on the average ambient groundwater temperature in Shenyang, China. A medium (27.5 °C) and a warm (33.0 °C) temperatures were chosen to allow the growth and good performance of denitrifers based on the optimum temperature (25–35 °C) (Karanasios et al. 2010). A comparison was made of denitrification rates among different water temperatures in Fig. 2.8a. The changes in NO_2-N and NH_4-N over reaction time at different water temperatures are shown in Fig. 2.8b, c.

Denitrification capacity increased as temperature increased (Fig. 2.8a). The denitrification rate at 33 °C was 3.19 times higher than at 15.0 °C and 1.59 times higher than at 27.5 °C, and the denitrification rate at 27.5 °C was 2.02 times higher than at 15.0 °C, when complete NO_3-N removal was attained (Fig. 2.8a). These results demonstrated that the HAD process was temperature dependent and sensitive to change in temperature. These two conclusions were consistent with the literature describing cellulose-based HD (Volokita et al. 1996b; Robertson et al. 2000) in that HD was the most important NO_3-N denitrification mechanism in the HAD process (see Sect. 2.3.2.5). High temperatures positively affect bacterial enzyme kinetics, specific growth rates and reaction rate constants. On the other hand, the higher water temperature, the easier the breakdown of the carbon chain

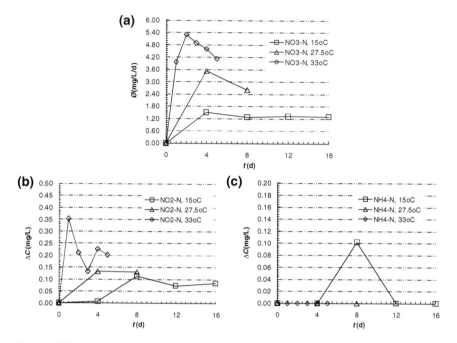

Fig. 2.8 Effect of water temperature on the performance of heterotrophic-autotrophic denitrification (HAD) over reaction time at 15, 27.5 and 33 °C: **a** denitrification rates; **b** variations of NO_2-N; **c** variations of NH_4-N

of pine bark by denitrifying microorganisms, and thus the greater available soluble fraction of organic carbon. Zhao et al. (2009) found the elevation of temperature in the range of 10–30 °C resulted in a comparable positive influence on HD.

The maximum concentration of NO_2-N (0.35 mg/L NO_2-N) was recorded at 30 °C compared with 0.13 mg/L NO_2-N at 27.5 °C and 0.11 mg/L NO_2-N at 15 °C (Fig. 2.8b). Apparently, the activities of nitrate reductase at the warm temperature were slightly higher than those at the other two temperatures. No variations in NH_4-N were found at 15, 27.5 and 30 °C respectively, except after 8 days at 15 °C (0.10 mg/L) (Fig. 2.8c). These results indicated that water temperature did not alter the HAD denitrification passways.

2.3.3.3 Effect of High-Concentration Nitrate

As discussed in Sect. 2.3.2.5, at the normal initial NO_3-N concentration of approximately 22.6 mg/L (around the acceptable China, India and Australia values for public health), the HAD was able to completely remove NO_3-N within 16 days. However, from the literature it's not known how high-concentration nitrate affects the HAD performance. Based on a literature survey, a high NO_3-N concentration of around 100 mg/L was set. A comparison was made of denitrification rates between

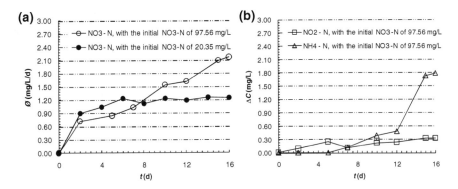

Fig. 2.9 Effect of high-concentration nitrate on the performance of heterotrophic-autotrophic denitrification (HAD) over reaction time: **a** denitrification rates at the initial NO₃-N of 97.58 and 20.35 mg/L; **b** variations of NO₂-N; **c** variations of NH₄-N

the two NO_3-N concentrations (20.35 and 97.58 mg/L) in Fig. 2.9a. The changes in NO_2-N and NH_4-N over reaction time at the high concentration (97.58 mg/L) are shown in Fig. 2.9b.

While the percent removal of NO_3-N decreased to 35.53 % at day 16 at the high NO_3-N of 97.58 mg/L (data not shown), the corresponding denitrification rate of 2.166 mg/L/d was much greater than that of 1.245 mg/L/d at the NO_3-N of 20.35 mg/L (Fig. 2.9a). These results suggested the high NO_3-N level was not seriously detrimental to the denitrifying biofilm and the NO_3-N substrate inhibition did not occur. It is thus clear that the denitrifiers had strong shock resistence ability. In an earlier study, Della Rocca et al. (2006) concluded that denitrifiers in the iron powder/cotton system could resist NO_3-N of 49.78 mg/L at hydraulic retention time (HRT) of approximate 4 days.

The variation in NO_2-N reached to 0.34 mg/L at day 16 at the high concentration of NO_3-N (Fig. 2.9b). To further increase percent removal as high as possible and decrease NO_2-N as low as possible at the high NO_3-N concentration incubation, increasing reaction time or decreasing mineral salts media volume might be required. It can be seen from Fig. 2.9b, starting from day 5, increasing amounts of NH_4-N were generated at the high NO_3-N concentration, with time. Since NH_4-N was not produced at the normal initial NO_3-N concentration (Fig. 2.2b), NH_4-N production at the high concentration implied that the release rate of carbon was lower than its consumption rate, and then the HAD denitrification was limited without sufficient supply of soluble organic carbon. Other researchers have also reached a similar conclusion (Su and Puls 2007).

2.3.3.4 Effect of Nitrite

Nitrite, an intermediate product in BD, is a highly toxic compound, whose toxicity may inhibit microbial enzyme activities. Nitrite accumulation was often observed

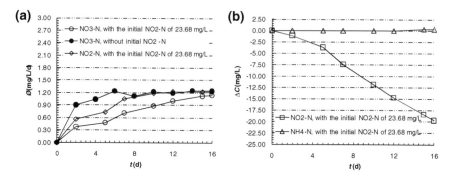

Fig. 2.10 Effect of nitrite on the performance of heterotrophic-autotrophic denitrification (HAD) over reaction time: **a** denitrification rates with and without the initial NO_2-N of 23.68 mg/L; **b** variations of NO_2-N; **c** variations of NH_4-N

in laboratory experiments and field applications by numerous researchers in literature (Piñar and Ramos 1998; Chang et al. 1999; Gómez et al. 2000; Smith et al. 2001; Lee and Rittmann 2002; Della Rocca et al. 2006). Nitrite reductase is rather sensitive to environmental conditions which include carbon source types, temperature, H_2 pressures, DO, phosphate, toxic compounds (e.g. heave metals, pesticides or their derivatives), electrical conductivity, and so forth. In addition, nitrite accumulation is strongly influenced by microbial species present. As discussed in Sect. 2.3.2.1, only ≤ 0.77 % of the NO_3-N was converted to NO_2-N at the initial NO_3-N of 20.35 mg/L. Even so, it is desirable to investigate the effect of nitrite on the HAD performance due to common occurrence of nitrite accumulation and complexity of a groundwater environment. 23.68 mg/L NO_2-N was added into a 2 L Schott bottle containing 20.48 mg/L NO_3-N. A comparison was made of NO_3-N denitrification rates between incubations containing 20.48 mg/L NO_3-N with and without initial NO_2-N (Fig. 2.10a), and the change in NO_2-N denitrification rates is shown in Fig. 2.10a. The changes in NO_2-N and NH_4-N over reaction time at the initial NO_2-N of 23.68 mg/L are shown in Fig. 2.10b.

The NO_3-N denitrification rates without initial NO_2-N were constantly greater than with it, but the difference between them became smaller and smaller (Fig. 2.10a). These indicated that: (1) nitrate reductase was temporarily, but significantly, inhibited by the toxicity of nitrite at the beginning of incubation; (2) nitrate reductase gradually adapted itself to the highly toxic environment with the added NO_2-N decreasing and reaction time increasing; (3) nitrate reductase refreshed its activities and syntheses at the end of incubation; and (4) there was a remarkable negative impact of nitrite on the HAD performance.

Unexpectedly, the NO_2-N denitrification rates were higher than the NO_3-N ones (Fig. 2.10a) and most of the added NO_2-N (19.68 mg/L) was reduced during the 16 day period at the initial NO_2-N of 23.68 mg/L (Fig. 2.10b). These suggested that: (1) the HAD was capable of simultaneously removing NO_3-N and NO_2-N; (2) the activities and syntheses of nitrite reductase was enhanced due to adequate nitrite substrate; (3) the activities of nitrite reductase were higher than those of nitrate reductase; (4) nitrite as an electron acceptor appeared to be superior to nitrate; and (5) soluble organic carbon released by pine bark met the needs of HD of nitrate and nitrite. Variations of NH_4-N were not found before day 15, and ascended to levels between 0.32 and 0.33 mg/L afterwards at the initial NO_2-N of 23.68 mg/L (Fig. 2.10b). Likewise, nitrite did not alter the HAD denitrification passways.

2.3.3.5 Effect of Ammonium

Although nitrate is most commonly associated with groundwater nitrogen contamination, combined contamination of nitrate and ammonium is also found, primarily due to the discharge of wastewater from sources such as septic systems and wastewater infiltration beds. It is possible that partial nitrate is reduced to ammonium by the HAD when denitrification conditions such as solution pH are changed. Ammonium is potentially toxic to aquatic organisms, particularly microorganisms, at high concentrations (Su and Puls 2007). Della Rocca et al. (2006) point out denitrifers were inhibited by ammonium (14.62 mg/L as NH_4-N). For these reasons, the effect of ammonium on the HAD performance was investigated. 21.22 mg/L NH_4-N was added into a 2L Schott bottle containing 21.46 mg/L NO_3-N. A comparison was made of denitrification rates between incubations containing 21.46 mg/L NO_3-N with 21.22 mg NH_4-N/L and without initial NH_4-N (Fig. 2.11a). The changes in NO_2-N and NH_4-N over reaction time at the initial NH_4-N of 21.22 mg/L are shown in Fig. 2.10b.

The denitrification rates of incubations without initial NH_4-N were higher, in the first 7 days of incubation, than those containing NH_4-N (Fig. 2.11a), indicating nitrate and nitrite reductase was significantly inhibited by ammonium at the beginning of incubation. In contrast the rates for incubations without NH_4-N were

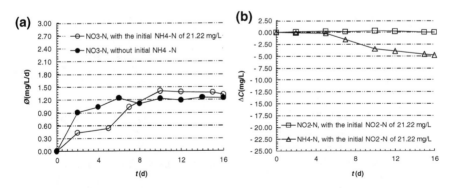

Fig. 2.11 Effect of ammonium on the performance of heterotrophic-autotrophic denitrification (HAD) over reaction time: **a** denitrification rates with and without the initial NH_4-N of 21.22 mg/L; **b** variations of NO_2-N; **c** variations of NH_4-N

lower than those containing NH_4-N after day 8 (Fig. 2.11a), indicating both reductase had adapted themself to the high ammonium (21.22 mg/L as NH_4-N).

Variations of NO_2-N were observed to vary between 0 and 0.32 mg/L (Fig. 2.11b), suggesting that complete denitrification was attained during the 16 day period in the presence of 21.22 mg/L NH_4-N. 4.85 mg/L NH_4-N was lost from the Schott bottle with initial NH_4-N in 16 days (Fig. 2.11b). Ammonia volatilization could not explain the NH_4-N loss as the pH values (≤ 8.45) (data not shown) were not above 9. The mixed bacteria used possibly contained anaerobic ammomium-oxidizing bacteria. Under anaerobic conditions, anaerobic ammomium-oxidizing bacteria could directly oxidize ammonium to nitrogen gas with nitrate and nitrite as the electron acceptors in the presence of both nitrate and ammonium (Eqs. 2.18, and 2.19) (Su and Puls 2004).

$$3NO_3^- + 5NH_4^+ \rightarrow 4N_2 + 9H_2O + 2H^+ \tag{2.18}$$

$$NO_2^- + NH_4^+ \rightarrow N_2 + 2H_2O \tag{2.19}$$

Little added NH_4-N (≤ 0.20 mg/L) was lost before day 5 (Fig. 2.11b). This behavior indicated adsorption onto pine bark and spongy iron was not a cause of the NH_4-N loss (4.85 mg/L).

2.3.3.6 Effects of Coexistent Inorganic Anions

Successful implementation of an in situ groundwater remediation process requires a thorough understanding of the effect of geochemical composition on its behavior. It is expected that some coexistent inorganic anions (such as SO_4^{2-}, SO_3^{2-} and BO_3^{3-}) will affect the behavior of the HAD process. However, their effects have not been well understood. A comparison was made of denitrification rates in incubations with and without coexistent inorganic anions in Fig. 2.12a. The changes in NO_2-N and NH_4-N over reaction time in the presence of different coexistent anions are shown in Fig. 2.12b, c.

Sulfate, sulfite and borate all had strongly similar inhibitory effects on the HAD, from days 2 to 16 (Fig. 2.12a). It's possible that the toxicity of sulfate, sulfite (more toxic) and borate (125, 104 and 513 mg/L respectively) inhibited the microbial activity of heterotrophic and autotrophic denitrifiers. The blockage of reactive sites on the surface of Fe^0 and its corrosion products by specific adsorption of the inner-sphere complex forming ligands (sulfate, sulfite and borate) (Su and Puls 2004) might be responsible for the decreased nitrate CR by spongy iron in the HAD. The denitrification rates with sulfate, sulfite and borate at day 16 were lower than without them (Fig. 2.12a), suggesting NO_3-N was not completely depleted due to the existence of sulfate, sulfite and borate. Robertson et al. (2007) reported sulfate and sulfite reduction consuming organic carbon (Eqs. 2.20, 2.21) did not occur until after NO_3-N depletion was complete. It can be deduced that sulfate and sulfite reduction did not take place in the 16 days.

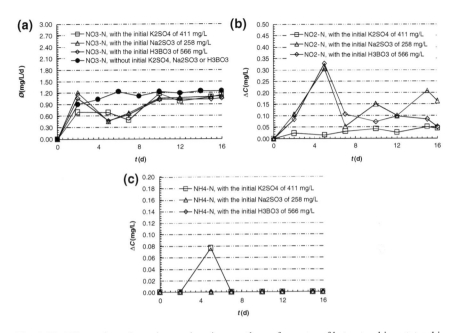

Fig. 2.12 Effects of coexistent inorganic anions on the performance of heterotrophic-autotrophic denitrification (HAD) over reaction time: **a** denitrification rates with the initial K_2SO_4 of 411 mg/L, the initial Na_2SO_3 of 258 mg/L, the initial H_3BO_3 of 566 mg/L, and without any of them; **b** variations of NO_2-N; **c** variations of NH_4-N

$$SO_4^{2-} + 2CH_2O \rightarrow 2HCO_3^- + H_2S \qquad (2.20)$$

$$2SO_3^{2-} + 3CH_2O + H^+ \rightarrow 3HCO_3^- + 2H_2S \qquad (2.21)$$

From trace to small amounts of NO_2-N (up to 0.33 mg/L) were detected in the presence of SO_4^{2-}, SO_3^{2-} and BO_3^{3-} (Fig. 2.12b), which was consistent with our studies in the above sections. NH_4-N was lower than 0.1 mg/L (Fig. 2.12c), so it was not a predominant denitrification product in the presence of the three coexistent inorganic anions.

The presence of SO_4^{2-}, SO_3^{2-} and BO_3^{3-} in groundwater will retard NO_3-N denitrification by the HAD process. Their negative effects should thus be taken into consideration when a PRB is designed for field application. Consequently, a greater mass of spongy iron and/or pine bark may be needed for sites where SO_4^{2-}, SO_3^{2-} and BO_3^{3-} are significant co-contaminants with nitrate in groundwater.

2.3.4 Long-term Performance of the Heterotrophic-Autotrophic Denitrification Process

To investigate the long-term performance of HAD, a HAD bottle was kept in incubation continuously for 105 days. When both NO_3-N and NO_2-N concentrations were no longer detected, NO_3-N was added to return the solutions to final concentrations of between 20.77–23.62 mg/L, however, neither pine bark nor spongy iron was supplemented.

The experimental results for the changes in NO_3-N, NO_2-N, NH_4-N and denitrification rate are illustrated in Fig. 2.13a–c. It was observed that complete NO_3-N removal was achieved rapidly in approximately 16 days after each NO_3-N addition (Fig. 2.13a), and almost equal denitrification rates (1.233–1.397 mg/L/d) were obtained when NO_3-N was completely removed at days 15, 34, 53, 70, 87, 105 (Fig. 2.13c). All the evidence from the present study indicated pine bark could provide sufficient organic carbon, spongy iron could steadily remove DO, and microbial activity maintained relatively constant during 105 days. Pine bark, rich in cellulose fibers, released organic carbon by means of cellulose-degrading bacteria, and simultaneously heterotrophic denitrifiers consumed organic carbon to finish their metabolism. Apparently, the net result of these two different processes was sufficient supply of organic carbon. Ovez et al. (2006) showed that the time

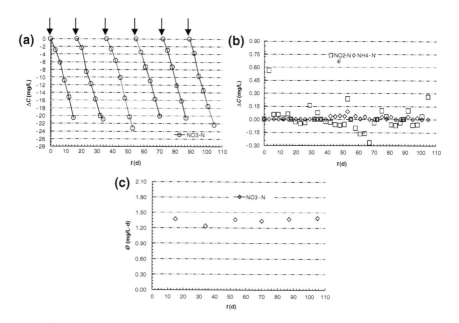

Fig. 2.13 Long-term performance of heterotrophic-autotrophic denitrification (HAD) over reaction time at the initial NO_3-N of 20.77–23.62 mg/L: **a** variations of NO_3-N; **b** variations of NO_2-N and NH_4-N; **c** variations of denitrification rates. Arrows indicate the addition of NO_3-N to the Schott bottle

(<10 days) for 100 % NO_3-N removals was always shorter in liquorice and giant reed reactors at the beginning of incubation and both of the reactors provided higher denitrification rates for a long-time (70 days), when compared to the HAD bottle in this study. Nevertheless, denitrification rates for liquorice and giant reed declined as the process continued because of the reduction in the usable organic carbon content of the substances. Volokita et al. (1996b) found that a time-dependent decay in denitrification rate was noticeable after several months of operation when newspaper served as a sole carbon source. Of the four cellulose-based organic substances, pine bark was capable of providing the most steady denitrification rates for 3.5 months. Pine bark was superior to liquorice, giant reed and newspaper. Only at day 3 was a significant, high and transient NO_2-N concentration (0.56 mg/L) generated (Fig. 2.13b). NO_2-N concentrations of > 0.3 mg/L generated represented 2.63 % of the 38 water samples collected. These phenomena suggested that no NO_2-N accumulated during the long-term incubation (105 days). A maximum of NH_4-N generated (0.09 mg/L) was observed at day 53 (Fig. 2.13b), which only accounted for 0.4 % of NO_3-N removed. This indicated that NO_3-N was not converted into NH_4-N during the long-term incubation.

2.3.5 Kinetics of NO₃-N Denitrification by the Heterotrophic-Autotrophic Denitrification Process

To study the kinetics of NO_3-N denitrification by the HAD process, three different initial NO_3-N concentrations (44.20, 20.35 and 11.15 mg/L) were incubated and the changes in NO_3-N over reaction time were monitored and compared with an incubation containing 20.35 mg/L NO_3-N at a DO concentration of 4.1 mg/L in the absence of pine bark, spongy iron and mixed bacteria (Fig. 2.14a-c).

From Fig. 2.14a-c, about 32, 16 and 8 days were needed to completely remove nitrate at the initial NO_3-N concentrations of 44.20, 20.35 and 11.15 mg/L respectively. The higher the initial NO_3-N concentration was, increased the time for complete nitrate removal. It can also be seen from Fig. 2.14a-c that NO_3-N decreased with increasing reaction time and the change in trends of NO_3-N concentration for all the three concentrations were similar. The results for a blank solution containing NO_3-N (20.35 mg/L) but no spongy iron, pine bark and mixed bacteria are included in Fig. 2.14b and show losses of NO_3-N to be insignificant in the absence of pine bark, spongy iron and mixed bacteria.

The above observation implied that NO_3-N denitrification under these conditions would comply with zero order reactions with respect to NO_3-N concentration. A zero order reaction with respect to NO_3-N concentration can be described by a differential equation as given in Eq. 2.22.

$$\frac{-d[NO_3^-]}{dt} = K \tag{2.22}$$

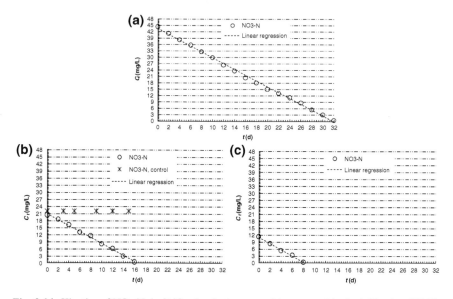

Fig. 2.14 Kinetics of NO$_3$-N denitrification by heterotrophic-autotrophic denitrification (HAD): **a** the initial NO$_3$-N of 44.20 mg/L; **b** the initial NO$_3$-N of 20.35 mg/L; **c** the initial NO$_3$-N of 11.15 mg/L. The control solution contained NO$_3$-N of 20.35 mg/L and DO of 4.1 mg/L in the absence of granulated spongy iron, pine bark and mixed bacteria

If Eq. 2.22 is integrated it gives an equation often called the integrated zero order rate law as given in Eq. 2.23.

$$[NO_3^-] = -Kt + [NO_3^-]_0 \qquad (2.23)$$

where K is the zero order reaction rate constant, whose negative is the observed slope of the regression line from plotting NO$_3$-N concentration against reaction time.

A linear regression of the zero order kinetic Eq. 2.22 was performed for each data set (Fig. 2.14a–c). The linear regression model was statistically significant ($R^2 > 0.9949$) in all cases (Eqs. 2.24, 2.25, and 2.26) providing strong evidence that the HAD denitrification was zero order with respect to NO$_3$-N concentration, which suggested that there were sufficient denitrifiers and (organic) carbon sources in the HAD process.

At the initial NO$_3$-N of ~45.2 mg/L:

$$C(mg/L) = -1.3654t(d) + 43.3647 \, (R^2 = 0.9978) \qquad (2.24)$$

At the initial NO$_3$-N of ~22.6 mg/L:

$$C(mg/L) = -1.2676t(d) + 20.8864 \, (R^2 = 0.9965) \qquad (2.25)$$

At the initial NO$_3$-N of ~11.3 mg/L:

$$C(\text{mg/L}) = -1.3333t(\text{d}) + 10.9350 \left(R^2 = 0.9949\right) \tag{2.26}$$

The values of K ranged from 1.2676 to 1.3654 mg/L/d (Eqs. 2.24, 2.25, and 2.26) with a mean of 1.3220 mg/L/d. Little variation in K among the three different initial NO_3-N concentrations indicated that altering the concentration of initial NO_3-N did not alter the rate of reaction, i.e., the amount of NO_3-N removed was positively proportional to the reaction time. It is therefore concluded that under the conditions of the experiment, the HAD denitrification was zero order with respect to NO_3-N concentration and the reaction rate constant (K) was independent of the initial NO_3-N concentrations within the applied concentration range.

The K, intimately related to environmental conditions (water temperature, pH, DO, etc.) and experimental parameters (denitrifier concentration, masses of spongy iron and pine bark, etc.), had great potential application prospects in designing the HAD process. For example, if influent and effluent NO_3-N concentrations are given, the hydraulic retention time will be determined most easily with the K.

2.3.6 Mechanisms for Dissolved Oxygen Removal and NO_3-N Denitrification in the Heterotrophic-Autotrophic Denitrification Process

The main mechanisms for DO removal and NO_3-N denitrification in the HAD process are illustrated in Fig. 2.15. Cellulose fiber-rich pine bark released organic carbon into mineral salts media by cellulose-degrading bacteria to support biological deoxygenation and HD. Spongy iron and aerobic heterotrophs rapidly removed DO respectively (Eqs. 2.8, 2.9), creating anaerobic environments to favor BD, chemical nitrate reduction and anaerobic iron corrosion. Simultaneously, biological deoxygenation produced carbon dioxide, favoring AD. 0.142 d and 1.905 d were needed to completely remove DO by granulated spongy iron-based CR and aerobic heterotrophs respectively. Granulated spongy iron-based CR played a dominant role in DO removal. Under anaerobic conditions, the corrosion of spongy iron was coupled with the reduction of water-derived protons, forming cathodic hydrogen (Eq. 2.13) to favor AD. At the same time, bacteria maybe took advantage of hydrogenase enzymes to accelerate the anodic dissolution of iron and hydrogen production to some extent when H^+ or H_2S was generated (Eqs. 2.18, 2.20, and 2.21).

Cathodic hydrogen and inorganic carbon (CO_2 and HCO_3^-) allowed autorotrophic denitrifiers to reduce nitrate to innocuous nitrogen gas (Eq. 2.17). 4.5–8.7 % of the intial NO_3-N was denitrified by AD in 16 days. At neutral or weak alkaline pH values, spongy iron chemically reduced nitrate. 6.2 % of the intial NO_3-N was reduced by granulated spongy iron-based CR in 16 days. Possible pathways of chemical nitrate reduction are shown in Eqs. 2.10, 2.11, and 2.12. Organic carbon released by pine bark acted both as an electron donor and as carbon and energy source. Heterotrophic denitrifiers utilized the organic carbon to reduce

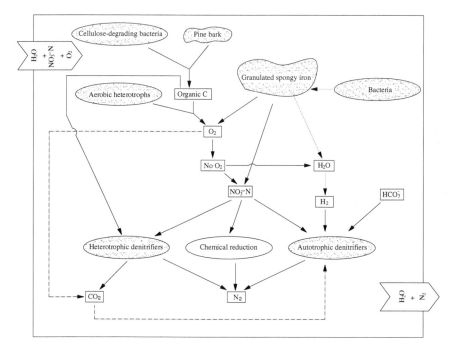

Fig. 2.15 The mechanisms for dissolved oxygen (DO) removal and NO_3-N denitrification occurring in the heterotrophic-autotrophic denitrification (HAD) process

nitrate and meanwhile generated carbon dioxide, which further favored AD (Eq. 2.14). 83.1 % of the intial NO_3-N was removed by HD in 16 days. HD was the most important mechanism for NO_3-N denitrification in the HAD. Pine bark providing a large amount of surface area served as a solid-carrier for biofilm formation. Generally, spongy iron has a porous internal structure and large surface area. The spongy iron used having a Brunauer-Emmett-Teller (BET) specific surface of 0.49 m^2/g served as another solid-carrier.

In short, the combination of cellulose-degrading bacteria, aerobic heterotrophs, granulated spongy iron-based CR, anaerobic iron corrosion, heterotrophic and autotrophic denitrifiers contributed to DO removal and NO_3-N denitrification in the HAD process.

2.3.7 Comparison of Denitrification Capacity Among Different Denitrification Approaches

Based on a literature survey, the comparison of maximum NO_3-N percent removal (MNPR) (%) and maximum denitrification rate (MDR) (mg/L/d) among different denitrification approaches is presented in Table 2.3. In spite of the similar MNPRs ($\geq$96.3 %) for all the processes, the differences in denitrification rate among these

Table 2.3 Comparison of denitrification capacity among different denitrification processes

Material	System description	Maximum NO_3-N percent removal (%)	Maximum denitrification rate (mg/L/d)	Water temperature (°C)	Denitrification mechanism	Reference
Fe powder (10 μm)	Batch study	100	780.48[a]	No[b]	CR	Liao et al. (2003)
Fe powder (100–200 mesh)	Batch study	~100	1290.86[a]	22	CR	Siantar et al. (1996)
Nanoscale Cu/Fe (5.0 %, w/w)	Batch study	100	280[a]	25	CR	Liou et al. (2005)
Iron grains (~0.5 mm)	Batch study	~100	1509.6[a]	No[b]	CR	Huang and Zhang (2004)
Acetate	Batch study	~100	0.55[a]	No[b]	HD	Devlin et al. (2000)
Cotton burr compost	Batch study	~100	6.67[a]	23 ± 1	HD	Su and Puls (2007)
Cornstalks	Batch study	No[b]	13.54 ± 1.74[a]	20 ± 2	HD	Greenan et al. (2006)
Cardboard fibers	Batch study	No[b]	5.18 ± 0.11[a]	20 ± 2	HD	Greenan et al. (2006)
Wood chips	Batch study	No[b]	2.19 ± 0.20[a]	20 ± 2	HD	Greenan et al. (2006)
Wood chips amended with soybean oil	Batch study	No[b]	3.75 ± 0.09[a]	20 ± 2	HD	Greenan et al. (2006)
Pure H_2	Batch study	96.3[a]	7.30[a]	30	AD	Smith et al. (1994)
Pure H_2 + CO_2	Batch study	~100	75.98	30 ± 1	AD	Vasiliadou et al. (2006)
Cathodic hydrogen + HCO_3^- +nanoscale Fe	Batch study	~100	46.67[a]	25	AD	Shin and Cha (2008)
Elemental sulfur + limestone	Batch study	100	No[b]	22	AD	Zhang and Lampe (1999)
Methanol + spongy Fe (0.075–0.425 mm)	Batch study	~100	4.2	27.5	HD + AD + CR	Huang et al. (2012)
Pine bark + spongy Fe (0.075–0.425 mm)	Batch study	~100	1.25	15	HD + AD + CR	This study

[a] Estimated or calculated based on the data given in the article; [b] No data

processes were very evident (Table 2.3). This is the consequence of the differences in denitrification mechanism and experimental condition. Even though the MDRs for the single CR processes were very high (Table 2.3), but the main problems of Fe^0-based CR were the release of ammonium and ferrous ions, and a requirement for a low acidic pH (or a pH buffer or a bimetallic catalyst), which limit their applicability. Although the MDRs for the single hydrogentrophic denitrification processes were ideal (Table 2.3) and hydrogentrophic denitrification could remove nitrate more cleanly with the production of fewer residual organics compared to HD, the explosion and safety concerns and the low solubility associated with pure H_2 have prevented widespread acceptance of hydrogenotrophic denitrification as a remediation technology. The agglomeration or clumping of nanoscale iron particles to each other or to the substrate surface has impeded field applications of nanoscale iron-based CR or hydrogenotrophic denitrification. The sulfur-autotrophic denitrification process has showed some drawbacks: the low solubility of sulfur compounds; the production of sulfate and nitrite; the use of limestone for pH adjustment (Zhang and Lampe 1999; Karanasios et al. 2010). These drawbacks limit its applicability. The MDR for the HAD process reported in this study was lower than the values given in Table 2.3 for the HD processes (except the acetate-supported HD process). Nevertheless, the HD processes had low or even no capacities of deoxygenation. When encountering DO, these processes would weaken or even lose their denitrification performances. In contrast, the HAD process was able to effectively remove DO (Fig. 2.1), which could maintain its stable denitrification rate. The MDR for the HAD process supported by methanol and spongy iron was acceptable, but maintaining contact between methanol and spongy iron would require a large amount of methanol in field applications in that methanol is water soluble and mobile, and spongy iron is neither. More important, methanol is considerably more toxic than nitrate, which makes its use restricted to the treatment of wastewater. Both disadvantages have hindered field applications of methanol-supported HD and HAD for in situ groundwater remediation.

2.4 Conclusions

Based on the research findings the main conclusions were as follows.

1. The HAD process's deoxygenation capacity depended on granulated spongy iron via CR and aerobic heterotrophs via aerobic respiration to remove DO, but granulated spongy iron played a dominant role in deoxygenation. 0.121 d, 0.142 d and 1.905 d were needed to completely remove DO by HAD, granulated spongy iron and mixed bacteria respectively.
2. The HAD process was potentially a feasible and effective approach for groundwater remediation. 19.92 mg/L NO_3-N was removed during the 16 day period. No NO_2-N accumulation occurred. No NH_4-N variations were found.
3. There existed symbiotic, synergistic and promotive effects of CR, HD and AD within the HAD. CR, HD and AD all contributed to the overall removal of

NO_3-N, but HD was the most important NO_3-N denitrification mechanism. After 16 days, NO_3-N removal was approximately 100, 6.2, 83.1, 4.5 % by HAD, CR, HD, AD, respectively.

4. The HAD denitrification was zero order with respect to NO_3-N concentration, and the reaction rate constant (K) of 1.3220 mg/L/d was independent of the initial NO_3-N concentrations within the applied concentration range (11.15–44.20 mg/L).

5. There was a minor effect of pine bark amount on the HAD denitrification from days 0 to 4, whereas there was a noticeable effect from days 4 to 16.

6. Denitrification capacity increased as temperature increased. The denitrification rate at 33 °C was 3.19 times higher than at 15.0 °C and 1.59 times higher than at 27.5 °C.

7. The high nitrate level (97.58 mg/L as NO_3-N) was not seriously detrimental to the denitrifying biofilm. The denitrifiers had strong shock resistence ability. The denitrification rate of 2.17 mg/L/d at the high NO_3-N concentration was much greater than that of 1.25 mg/L/d at the normal NO_3-N concentration (20.35 mg/L).

8. Nitrite (23.68 mg/L as NO_3-N) had a remarkable negative impact on the HAD performance. Nitrate reductase was temporarily, but significantly, inhibited by the toxicity of nitrite at the beginning of incubation, and then it refreshed its activities and syntheses at the end.

9. High ammonium (21.22 mg/L NH_4-N) significantly inhibited nitrate and nitrite reductase at the beginning of incubation. Afterwards both reductase adapted themself to the high ammonium.

10. Coexistent inorganic anions (125 mg/L sulfate, 104 mg/L sulfite and 513 mg/L borate) all had strongly similar inhibitory effects on the HAD performance, from days 2 to 16. NO_3-N was not completely depleted at day 16 due to their existence.

11. The HAD approach was capable of providing steady denitrification rate (1.233–1.397 mg/L/d for 3.5 months. Pine bark could provide sufficient organic carbon, spongy iron could steadily remove DO, and microbial activity maintained relatively constant. Almost no NO_2-N accumulated and no NH_4-N was generated during the long-term incubation.

References

Ahn SC, Oh SY, Cha DK (2008) Enhanced reduction of nitrate by zero-valent iron at elevated temperatures. J Hazard Mater 156:17–22

APHA, AWWA, WPCF (1992) Standard methods for the examination of water and wastewater, 18th edn. American Public Health Association (APHA), American Water Works Association (AWWA) & Water Pollution Control Federation (WPCF), Washington, DC

Beaubien A, Hu Y, Bellahcen D, Urbain V, Chang J (1995) Monitoring metabolic activity of denitrification processes using gas production measurements. Water Res 29(10):2269–2274

Betlach MR, Tiedje JM (1981) Kinetic explanation for accumulation of nitrite, nitric-oxide, and nitrous-oxide during bacterial denitrification. Appl Environ Microbiol 42(6):1074–1084

Chang C, Tseng S, Huang H (1999) Hydrogenotrophic denitrification with immobilized *Alcaligenes eutrophus* for drinking water treatment. Bioresour Technol 69:53–58

Cheng IF, Muftikian R, Fernando Q, Korte N (1997) Reduction of nitrate to ammonia by zero-valent iron. Chemosphere 35:2689–2695

Choe S, Chang YY, Hwang KY, Khim J (2000) Kinetics of reductive denitrification by nanoscale zero-valent iron. Chemosphere 41:1307–1311

Choe SH, Ljestrand HM, Khim J (2004) Nitrate reduction by zero-valent iron under different pH regimes. Appl Geochem 19:335–342

Della Rocca C, Belgiorno V, Meric S (2005a) Innovative heterotrophic/autotrophic denitrification (HAD) of drinking water: Effect of ZVI on nitrate removal. A265-A270 in Proceedings of the 9th international conference on environmental science and technology. Rhodes island, Greece, 1–3 Sept 2005

Della Rocca C, Belgiorno V, Meric S (2005b) Cotton-supported heterotrophic denitrification of nitrate-rich drinking water with a sand filtration post-treatment. Water SA 31:229–236

Della Rocca C, Belgiorno V, Meric S (2006) An heterotrophic/autotrophic denitrification (HAD) approach for nitrate removal from drinking water. Process Biochem 41:1022–1028

Della Rocca C, Belgiorno V, Meriç S (2007a) Overview of in situ applicable nitrate removal processes. Desalination 204:46–62

Della Rocca C, Belgiorno V, Meric S (2007b) Heterotrophic/autotrophic denitrification (HAD) of drinking water: Prospective use for permeable reactive barrier. Desalination 210:194–204

Devlin JF, Eedy R, Butler BJ (2000) The effects of electron donor and granular iron on nitrate transformation rates in sediments from a municipal water supply aquifer. J Contam Hydrol 46:81–97

Elefsiniotis P, Li D (2006) The effect of temperature and carbon source on denitrification using volatile fatty acids. Biochem Eng J 28:148–155

Greenan CM, Moorman TB, Kaspar TC, Parkin TB, Jaynes DB (2006) Comparing carbon substrates for denitrification of subsurface drainage water. J Environ Qual 35:824–829

Gómez MA, González-López J, Hontoria-García E (2000) Influence of carbon source on nitrate removal of contaminated groundwater in a denitrifying submerged filter. J Hazard Mater 80:69–80

Gómez MA, Hontoria E, González-López J (2002) Effect of dissolved oxygen concentration on nitrate removal from groundwater using a denitrifying submerged filter. J Hazard Mater 90:267–278

Haugen KS, Semmens MJ, Novak PJ (2002) A novel in situ technology for the treatment of nitrate contaminated groundwater. Water Res 36:3497–3506

Huang CP, Wang HW, Chiu PC (1998) Nitrate reduction by metallic iron. Water Res 32:2257–2264

Huang G, Fallowfield H, Guan H, Liu F (2012) Remediation of nitrate-nitrogen contaminated groundwater by a heterotrophic-autotrophic denitrification (HAD) approach in an aerobic environment. Water Air Soil Pollut 223(7):4029–4038

Huang YH, Zhang TC (2004) Effects of low pH on nitrate reduction by iron powder. Water Res 38:2631–2642

Hunter WJ (2003) Accumulation of nitrite in denitrifying barriers when phosphate is limiting. J Contam Hydrol 66:79–91

Karanasios KA, Vasiliadou IA, Pavlou S, Vayenasa DV (2010) Hydrogenotrophic denitrification of potable water: a review. J Hazard Mater 180:20–37

Kielemoes J, De Boever P, Verstraete W (2000) Influence of denitrification on the corrosion of iron and stainless steel powder. Environ Sci Technol 34:663–671

Kim H, Seagren EA, Davis AP (2003) Engineered bioretention for removal of nitrate from stormwater runoff. Water Environ Res 75:355–367

Lee K, Rittmann BE (2002) Applying a novel autohydrogenotrophic hollow-fiber membrane biofilm reactor for denitrification of drinking water. Water Res 36:2040–2052

Liao C-H, Kang S-F, Hsu Y-W (2003) Zero-valent iron reduction of nitrate in the presence of ultraviolet light, organic matter and hydrogen peroxide. Water Res 37:4109–4118

Liou YH, Lo SL, Lin CJ, Kuan WH, Weng SC (2005) Chemical reduction of an unbuffered nitrate solution using catalyzed and uncatalyzed nanoscale iron particles. J Hazard Mater 127:102–110

Ovez B, Ozgen S, Yuksel M (2006) Biological denitrification in drinking water using Glycyrrhiza glabra and Arunda donax as the carbon source. Process Biochem 41:1539–1544

Piñar G, Ramos JL (1998) Recombinant *Klebsiella oxytoca* strains with improved efficiency in removal of high nitrate loads. Appl Environ Microbiol 64:5016–5019

Robertson WD, Blowes DW, Ptacek CJ, Cherry JA (2000) Long-term performance of in situ reactive barriers for nitrate remediation. Ground Water 38:689–695

Robertson WD, Ptacek CJ, Brown SJ (2007) Geochemical and hydrogeological impacts of a wood particle barrier treating nitrate and perchlorate in ground water. Ground Water Monit Remediat 27(2):85–95

Robertson WD, Vogan JL, Lombardo PS (2008) Nitrate removal rates in a 15-year-old permeable reactive barrier treating septic system nitrate. Ground Water Monit Remediat 28:65–72

Rodríguez-Maroto JM, Garcia-Herruzo F, Garcia-Rubio A, Sampaio LA (2009) Kinetics of the chemical reduction of nitrate by zero-valent iron. Chemosphere 74:804–809

Saliling WJB, Westerman PW, Losordo TM (2007) Wood chips and wheat straw as alternative biofilter media for denitrification reactors treating aquaculture and other wastewaters with high nitrate concentrations. Aquacult Eng 37:222–233

Schnobrich MR, Chaplin BP, Semmens MJ, Novak PJ (2007) Stimulating hydrogenotrophic denitrification in simulated groundwater containing high dissolved oxygen and nitrate concentrations. Water Res 41:1869–1876

Shin KH, Cha DK (2008) Microbial reduction of nitrate in the presence of nanoscale zero-valent iron. Chemosphere 72:257–262

Siantar DP, Schreier CG, Chou CS, Reinhard M (1996) Treatment of 1,2-dibromo-3-chloropropane and nitrate-contaminated water with zero-valent iron or hydrogen/palladium catalysts. Water Res 30:2315–2322

Smith RL, Ceazan ML, Brooks MH (1994) Autotrophic, hydrogen-oxidizing, denitrifying bacteria in groundwater, potential agents for bioremediation of nitrate contamination. Appl Environ Microbiol 60:1949–1955

Smith RL, Miller DN, Brooks MH, Widdowson MA, Killingstad MW (2001) In situ stimulation of groundwater denitrification with formate to remediate nitrate contamination. Environ Sci Technol 35:196–203

Soares MIM, Abeliovich A (1998) Wheat straw as substrate for water denitrification. Water Res 32:3790–3794

Soares MIM, Brenner A, Yevzori A, Messalem R, Leroux Y, Abeliovich A (2000) Denitrification of groundwater: pilot-plant testing of cotton-packed bioreactor and post-microfiltration. Wat Sci Technol 42:353–359

Straub KL, Buchholz-Cleven BEE (1998) Enumeration and detection of anaerobic ferrous iron-oxidizing, nitrate-reducing bacteria from diverse European sediments. Appl Environ Microbiol 64:4846–4856

Su C, Puls RW (2004) Nitrate Reduction by Zerovalent Iron: effects of formate, oxalate, citrate, chloride, sulfate, borate, and phosphate. Environ Sci Technol 38:2715–2720

Su C, Puls RW (2007) Removal of added nitrate in cotton burr compost, mulch compost, and peat: mechanisms and potential use for groundwater nitrate remediation. Chemosphere 66:91–98

Till BA, Weathers LJ, Alvarez PJJ (1998) Fe(0)-supported autotrophic denitrification. Environ Sci Technol 32:634–639

Tsai YJ, Chou FC, Cheng T-C (2009) Coupled acidification and ultrasound with iron enhances nitrate reduction. J Hazard Mater 163:743–747

Vasiliadou IA, Pavlou S, Vayenas DV (2006) A kinetic study of hydrogenotrophic denitrification. Process Biochem 41:1401–1408

Volokita M, Abeliovich A, Soares MIM (1996a) Denitrification of groundwater using cotton as energy source. Wat Sci Technol 34(1–2):379–385

Volokita M, Belkin S, Abeliovich A, Soares MIM (1996b) Biological denitrification of drinking water using newspaper. Water Res 30:965–971

Westerhoff P, James J (2003) Nitrate removal in zero-valent iron packed columns. Water Res 37:1818–1830

Yang GCC, Lee HL (2005) Chemical reduction of nitrate by nanosized iron: kinetics and pathways. Water Res 39:884–894

Weber KA, Pollock J, Cole KA, O'Connor SM, Achenbach LA, Coates JD (2006) Anaerobic nitrate-dependent iron(II) bio-oxidation by a novel lithoautotrophic betaproteobacterium, strain 2002. Appl Environ Microbiol 72:686–694

Yin SX, Chen D, Chen LM, Edis R (2002) Dissimilatory nitrate reduction to ammonium and responsible microorganisms in two Chinese and Australian paddy soils. Soil Biol Biochem 34:1131–1137

Zhang TC, Lampe DG (1999) Sulfur: limestone autotrophic denitrification processes for treatment of nitrate-contaminated water: Batch experiments. Water Res 33:599–608

Zhang Y, Zhong F, Xi S, Wang X, Li J (2009) Autohydrogenotrophic denitrification of drinking water using a polyvinyl chloride hollow fiber membrane biofilm reactor. J Hazard Mater 170:203–209

Zhao X, Meng XL, Wang JL (2009) Biological denitrification of drinking water using biodegradable polymer. Int J Environ Pollut 38:328–338

Chapter 3
Heterotrophic-Autotrophic Denitrification Permeable Reactive Barriers

Abstract Two potential heterotrophic-autotrophic denitrification permeable reactive barriers (HAD PRBs) were evaluated to remediate groundwater *in situ*. The first HAD PRB (Column 1) was packed with a mixture of spongy iron, pine bark and sand between 5 and 145 cm from bottom. The second HAD PRB (Column 2) was filled with a spongy iron and sand mixture layer between 5 and 35 cm from bottom, and a pine bark layer between 35 and 145 cm from bottom. The results showed that during operation over the 45 pore volumes, the influent NO_3-N concentration of ≤ 100 mg/L was mostly denitrified in Columns 1 and 2 at the flow rates of ≤ 0.30 m/d. The high NO_3-N percent removals (97–100 %) for both columns were achieved at hydraulic retention times ranging from 8.75 to 17.51 d. Most of the influent NO_3-N was removed in the first 25 cm at the low (23 mg/L) and middle (46 mg/L) NO_3-N concentrations and in the first 65 cm at the high concentration (104 mg/L) by Columns 1 and 2. Packing structure had a negligible effect on the performance of the two columns. Both HAD PRBs were highly feasible and effective in in situ groundwater remediation.

Keywords In situ groundwater remediation · Heterotrophic-autotrophic denitrification (HAD) · Permeable reactive barriers (PRBs) · Spongy iron · Pine bark · Pore volume · NO_3-N percent removal · Flow rates · Hydraulic retention times · Packing structure

Abbreviations

AD	Autotrophic denitrification
BD	Biological denitrification
BET	Brunauer-Emmett-Teller
CR	Chemical reduction
DO	Dissolved oxygen
HAD	Heterotrophic-autotrophic denitrification
HD	Heterotrophic denitrification
HRT	Hydraulic retention time
MNPR	Maximum NO_3-N percent removal

F. Liu et al., *Study on Heterotrophic-Autotrophic Denitrification Permeable Reactive Barriers (HAD PRBs) for In Situ Groundwater Remediation*, SpringerBriefs in Water Science and Technology, DOI: 10.1007/978-3-642-38154-6_3, © The Author(s) 2014

MVNLR   Maximum volumetric NO_3-N loading rate
PRB   Permeable reactive barrier
PV   Pore volume
SD   Standard deviation
TOC   Total organic carbon
VNLR   Volumetric NO_3-N loading rate
ZVI   Zero-valent iron

3.1 Introduction

Nitrate contamination of groundwater is a critical concern in China and through the world. Our previous batch studies have provided strong evidence that the heterotrophic-autotrophic denitrification (HAD) supported by granulated spongy iron and pine bark is an effective and feasible approach for nitrate removal from groundwater. In the batch studies, an incubation bottle is regarded as a closed system. This means: (1) mixed bacteria, spongy iron and pine bark are mixed well in a bottle; (2) the bacteria, the two media, and total organic carbon (TOC) in water are not lost from the bottle; and (3) bacterial biomass can accumulate with time. However, when designing and constructing a HAD permeable reactive barrier (PRB), the packing structure of spongy iron and pine bark media should be taken into account. Both media can be mixed or layered in a PRB. A mixed PRB and a layered PRB both will have abilities to chemically and biologically reduce nitrate-nitrogen (NO_3-N) due to the existence of spongy iron and aerobic heterotrophs in the front part of both PRBs, which can create an anaerobic environment for biological denitrification (BD). Unfortunately, the effect of packing structure on a HAD PRB performance is unknown. Moreover, in field applications, TOC and bacteria will easily be washed out with the effluent. Therefore it is unclear how the results from the batch studies apply to the field, where biomass and water quality indices (such as NO_3-N, nitrite-nitrogen (NO_2-N), ammonium-nitrogen (NH_4-N), dissolved oxygen (DO)) vary temporally and spatially and hydrogeological conditions (such as groundwater velocity and porosity) are characterized.

Compared with a batch test a continuous flow column experiment is by far a better way of observing variations of water quality indices, simulating a realistic groundwater environment, and identifying operating parameters. Therefore, based on batch studies, further studies using a continuous flow column system were needed prior to field applications of the HAD process. Several column studies have been conducted by others to explore the performances of BD and chemical nitrate reduction processes, which were favoured by sulfur, limestone, wood chips, wheat straw and scrap iron (Westerhoff and James 2003; Darbi et al. 2003; Della Rocca et al. 2006; Saliling et al. 2007; Moon et al. 2008). These primarily focused on evaluating short-term and long-term denitrification capacity, investigating the profiles of ions, determining optimal operating parameters and discovering problems associating with the use of these media. Although all of these column studies are very important, none of these combined HAD and PRB conceptually to treat nitrate contaminated groundwater in situ. Recent research has paid particular attention to nitrate removal by a PRB that intercepts and reduces nitrate in the groundwater plume. The potential PRB reactive media include zero-valent iron (ZVI) (Huang et al. 1998; Alowitz and Scherer 2002; Yang and Lee 2005; Ahn et al. 2008) and carbonaceous solid materials (Robertson and Cherry 1995; Schipper and Vojvodić-Vuković 1998; Soares and Abeliovich 1998; Robertson et al. 2000; Della Rocca et al. 2005). However, granulated spongy iron and pine bark as PRB materials have been neglected.

In this study, two HAD PRBs that were packed with granulated spongy iron and pine bark were proposed to remediate groundwater in situ in an aerobic environment (Fig. 3.1). The objectives of this study, using column experiments, were to: (1) explore the feasibility and efficiency of nitrate-nitrogen removal by the two HAD PRBs; (2) observe temporal and spatial variations of water quality indices (NO_3-N, NO_2-N, NH_4-N, DO, pH and TOC); (3) verify the effects of hydraulic retention time (HRT), volumetric NO_3-N loading rate (VNLR) and packing structure on the HAD PRBs; (4) compare denitrification capacity of the HAD PRBs with that of other denitrification processes in literature; and (5) provide reliable information on the HAD PRB processes for future field applications.

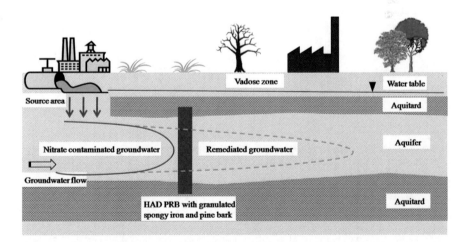

Fig. 3.1 Basic concept of the heterotrophic-autotrophic denitrification permeable reactive barrier (HAD PRB). As nitrate flows with the groundwater, it passes through the HAD PRB with granulated spongy iron and pine bark as reactive media

3.2 Materials and Methods

3.2.1 Materials and Chemicals

Granulated spongy iron (Table 3.1) was obtained from Kaibiyuan Co., Beijing, China. Pine bark was purchased from a local nursery store in Beijing, China. Moist sub-surface soil (0.3 m depth from surface) was taken from a pristine and humic-acid-rich area on the campus of Beijing Language and Culture University, Beijing, China. Gravel and sand were from a quarry in Beijing, China. Distilled water was used to prepare reagent solutions. Tap water (Table 3.2) spiked with $NaNO_3$, K_2HPO_4 and $NaHCO_3$ acted as synthetic groundwater. Unless otherwise indicated, all chemicals used were analytical reagent grade as received.

3.2.2 Inoculation

Initial enrichment culture was undertaken using a 20 L wide-mouth glass bottle to which was added: (1) 125 g of soil (0.15–0.45 mm); (2) 125 g of pine bark (0.45–2.0 mm); (3) 125 g of granulated spongy iron (0.15–2.0 mm); (4) 20 L of tap water; (5) 452 mg of NO_3-N; (6) 60 mg of HPO_4-P; and (7) 7 g of $NaHCO_3$. The N: P weight ratio of was 7.53: 1 and the initial DO was 5.5 mg/L. The bottle was sealed to create anaerobic conditions and covered with aluminum foil to shade out photosynthesizers (Rust et al. 2002). The bottle was incubated at 30 C without shaking. Following initial enrichment the denitrifying population was maintained by successively subculturing. When measured NO_3-N concentration was lower than 2.26 mg/L, a bacterial suspension was transferred into fresh tap water spiked with 22.6 mg/L of NO_3-N, 3 mg/L of HPO_4-P and 350 mg/L of $NaHCO_3$, at a bacterial suspension: tap water volume ratio of 1: 9 and a spongy iron: pine bark weight ratio of 1: 1.

3.2.3 Experimental Setup and Operation

Laboratory HAD PRB column experiments were conducted using two parallel columns made of plexiglass (150 cm in length, 20.6 cm internal diameter; Fig. 3.2).

Table 3.1 Element components of granulated spongy iron[a]

Fe^0/%	C/%	S/%	P/%	Mn/%	Ni/%	Cr/%	Cu/%	Al/%
60.60	0.78	0.06	0.04	0.29	0.02	0.02	0.02	0.26

[a] Data are provided by Metalllurgical Experimental Center, School of Metallurgical and Eco-logical Engineering, University of Science and Technology Beijing, China

Table 3.2 Average water chemistry of tap water used for study (unit: mg/L except pH)

Constituents	NO_3-N	NO_2-N	NH_4-N	F^-	Cl^-	SO_4^{2-}	HCO_3^-
Concentration	1.74	0.01	ND[a]	0.32	20.84	46.35	125.5
Constituents	Na^+	K^+	Ca^{2+}	Mg^{2+}	DO	pH	
Concentration	13.61	1.57	48.53	28.07	5–9	6.5–7.5	

[a] ND represents no detection

A lower support layer of 5 cm of clean gravel (2–5 mm) was packed at their bottom between 0 and 5 cm, and a 5 cm gravel cap was placed at their top between 145 and 150 cm. The first column (Column 1) was packed with a mixture of spongy iron, pine bark and sand between 5 and 145 cm (Fig. 3.2; Tables 3.3, 3.4). The second column (Column 2) was filled with a spongy iron and sand mixture layer between 5 and 35 cm, and a pine bark layer between 35 and 145 cm (Fig. 3.2; Tables 3.3, 3.4). The sand used was intended to improve sustainability of permeability. The synthetic groundwater was fed to the bottom of the two columns through 3.17 mm clear vinyl tubing. Similarly, vinyl tubing was also used to carry effluent and evolved gas away from the top of the columns (at 150 cm) for disposal (Fig. 3.2). The gas from the columns passed through a water seal in an effort to minimize the entry of air into each column. The vinyl tubing was cleaned with a 70 % ethanol solution at least once every 2 weeks to minimize biofilm and solids buildup inside the influent, effluent and gas lines. An adjustable multiport peristaltic pump set (BT 100-1F drive, DG-4 pump head, Baoding Longer Precision Pump Co., Ltd., Baoding, China) provided the fluid flow to the two columns in an upflow mode. A 20 L glass bottle served as the synthetic groundwater reservoir for each column. The bottle had the capacity for at least a three-day supply of the synthetic groundwater. Both columns were loaded continuously at variable operating conditions in 9 phases (Table 3.5) each for 5 pore volumes (PVs), in sequence. Each column was equipped with four sampling ports localized at 25, 65, 105, and 125 cm from the influent end and one gas port at the top (Fig. 3.2). The columns were covered with aluminum foil, and operated at room temperature (23 ± 5 °C).

3.2.4 Analytical Methods and Instruments

Samples were collected at the influent, effluent and the four sampling ports, and analysed for NO_3-N, NO_2-N, NH_4-N, TOC, pH, DO and water temperature. NO_3-N, NO_2-N and NH_4-N concentrations were determined using a ultraviolet-visible spectrophotometer (Hewlett Packard, Model 8453, USA). NO_3-N was measured by ultraviolet spectrophotometric method at 220 and 275 nm. NO_2-N was measured by N-(1-naphthyl)-ethylenediamine dihydrochloride colorimetric method at 540 nm. NH_4-N was measured by Nessler's reagent colorimetric method at 410 nm. The lower detection limits for NO_3-N, NO_2-N and NH_4-N were 0.08, 0.003 and 0.02 mg N/L respectively. TOC was determined with a TOC analyzer (Shimadzu, Model 3201, Japan) by measuring the difference between the total

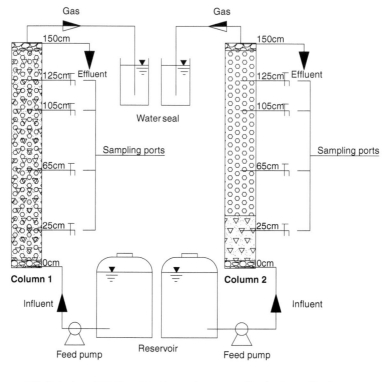

Fig. 3.2 Schematic diagram of the heterotrophic-autotrophic denitrification permeable reactive barrier (HAD PRB) columns (150 × 20.6 cm I. D. for each column) showing packed materials, flow direction and location of sampling ports used in this study

Table 3.3 Physical parameters of the heterotrophic-autotrophic denitrification permeable reactive barrier (HAD PRB) columns used for study

HAD PRB column	Spongy iron		Pine bark		Sand		Spongy iron to pine bark ratio (kg: kg)	Pore volume (L)	Average porosity (%)
	Mass (kg)	Particle size (mm)	Mass (kg)	Particle size (mm)	Mass (kg)	Particle size (mm)			
Column 1	7.99	0.15–2.00	7.13	0.45–2.0	7.40	0.45–2.0	1.1:1	28.73	62.95
Column 2	7.99	0.15–2.00	7.13	2.0–11.0	7.40	0.45–2.0	1.1:1	25.54	55.97

carbon (burning at 680 °C) and inorganic carbon (acidification with 2 M HCl). pH was determined using a digital pH meter (Sartorius, Model PB-10, Germany). DO and water temperature were monitored employing a portable DO meter (Hach, Model HQ30d, USA). Surface areas of granulated spongy iron, pine bark and sand were measured using a 5-point Brunauer-Emmett-Teller (BET) gas adsorption isotherm with N_2 gas on a surface area analyzer (Micromeritics, ASAP 2020, America).

Table 3.4 The characteristics of the packing media used for study

Packing media	NO$_3$-N (mg/g)	NO$_2$-N (mg/g)	NH$_4$-N (mg/g)	TOC (mg/g)	BET[d] (m^2/g)	Internal pore volume (cm^3/g)	Average pore diameter (Å)
Sand	ND[a]	ND	ND	0.75	2.10	0.0066	126.42
Spongy iron	ND	ND	ND	7.39	0.49	0.0012	101.87
Pine bark[b]	0.01	ND	3.92	31.53	0.79	0.0033	166.56
Pine bark[c]	0.01	ND	3.92	31.53	0.46	0.0018	159.42

[a] ND represents no detection; [b] Particle size: 0.45–2.0 mm; [c] Particle size: 2.0–11.0 mm; [d] BET: Brunauer-Emmett-Teller

3.2.5 Data Processing Methods

NO$_3$-N percent removal (Eq. 3.1)

$$\sigma = (C_{in} - C_{out})/C_{in} \times 100 \ \% \qquad (3.1)$$

where, σ is NO$_3$-N percent removal (%); C_{in} is influent NO$_3$-N concentration (mg/L); C_{out} is effluent NO$_3$-N concentration (mg/L)

Variation of pH (Eq. 3.2)

$$\Delta pH = pH_{out} - pH_{in} \qquad (3.2)$$

where, ΔpH is variation of pH; pH_{in} is influent pH; pH_{out} is effluent pH

Volumetric NO$_3$-N loading rate (Eq. 3.3)

$$L = F \times S \times C_{in}/V_b \qquad (3.3)$$

where, L is VNLR (g/m^3/d); F is linear flow rate (m/d); S is effective cross-sectional flow area of both columns (m^2); V_b is effective column volume (m^3)

Hydraulic retention time (Eq. 3.4)

$$\text{HRT} = V_b/(F \times S) \qquad (3.4)$$

where, HRT is hydraulic retention time (d)

3.3 Results and Discussion

3.3.1 Start-up

To start up the two HAD PRB columns, the inoculum accounting for 45 % of the pore volume was initially introduced into each column. The remaining volume in each column was filled with tap water, which was enriched with NO$_3$-N (41 mg/L), HPO$_4$-P (5.5 mg/L), NaHCO$_3$ (636 mg/L) and DO (8.2 mg/L). To enable aerobic heterotrophs and denitrifying bacteria to accumulate and form biofilms on the spongy iron and pine bark media, the contents of the columns were repeatedly

Table 3.5 Operating conditions of the two columns in different phases

Phase	I	II	III	IV	V	VI	VII	VIII	IX
Influent NO_3-N (mg/L)	22.18–24.34	21.79–24.42	22.69–24.35	44.23–49.69	47.86–51.25	46.14–49.34	102.12–104.87	101.38–104.72	102.55–104.23
Flow rate (m/d)	0.15	0.22	0.30	0.15	0.22	0.30	0.15	0.22	0.30

recirculated over an eight-day period. After 8 days, 80 % of the tap water was wasted when NO_3-N concentration was lower than 2.26 mg/L, and then replaced with the same amount of fresh tap water with $NaNO_3$, K_2HPO_4, $NaHCO_3$ and DO, in order to avoid nutrient limitation of biofilm growth. At day 20, the attached growth was considered substantial enough to cease the recirculation since >90 % of NO_3-N removal was achieved and no NO_2-N accumulation was observed in both columns (data not shown).

3.3.2 Temporal Variations of Water Quality Indices

When the bacterial consortium had successfully attached to the spongy iron and pine bark particles, the two parallel HAD PRB columns were fed continuously with the synthetic groundwater in 9 phases for a total of 45 PVs. This was conducted to observe the variations of NO_3-N, NO_2-N, NH_4-N, DO, pH, and TOC over time under variable operating conditions of influent NO_3-N concentration and flow rate. The experimental results are illustrated in Figs. 3.3, 3.4, 3.5, 3.6, 3.7, and 3.8; the dashed vertical lines delineate times when the influent NO_3-N concentration and/or the flow rate was changed. In these figures, number of PVs represents the ratio of the accumulated water volume over time to the pore volume of the reactive media.

3.3.2.1 Nitrate Variations

During operation over the initial 15 PVs, the NO_3-N concentration (mean ± standard deviation (SD)) for the column influent was 23.04 ± 0.81 mg/L for Column 1, and 22.66 ± 0.84 mg/L for Column 2; the effluent concentration of NO_3-N in Column 1 was 0.64 ± 0.35 mg/L, and in Column 2 was 0.61 ± 0.34 mg/L (Fig. 3.3a, b; Phases I, II, and III). During the second 15 PVs, the Column 1 NO_3-N concentration in the influent and effluent was 48.10 ± 1.91 and 1.46 ± 0.71 mg/L respectively, and the Column 2 corresponding values were 47.92 ± 1.61 and 1.29 ± 0.40 mg/L (Fig. 3.3a, b; Phases IV, V, and VI). During the third 15 PVs, the influent and effluent concentrations of NO_3-N for Column 1 were maintained at 103.34 ± 1.13 and 2.05 ± 0.73 mg/L; the corresponding values for Column 2 were kept at 103.09 ± 1.06 and 2.14 ± 0.79 mg/L (Fig. 3.3a, b; Phases VII, VIII, and IX). During operation over the 45 PVs, the maximum effluent NO_3-N concentrations of 3.42 and 3.22 mg/L were respectively observed at PV 40 for Column 1 and at PV 45 for Column 2. The influent NO_3-N concentration of up to 104 mg/L was mostly denitrified in Columns 1 and 2 (NO_3-N percent removals of >92 %) at the flow rates of up to 0.30 m/d. It can be seen from these data that both HAD PRB columns were highly effective in NO_3-N removal at the given influent NO_3-N concentrations and flow rates. It can be also seen that there was little difference in denitrification performance between columns 1 and 2.

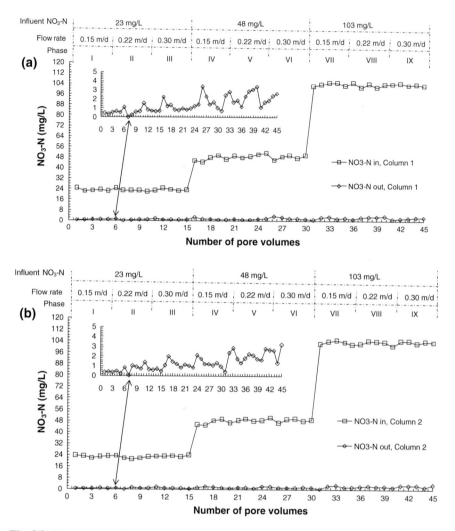

Fig. 3.3 Variations in NO₃-N over time under variable operating conditions in **a** Column 1 and **b** Column 2. *Vertical dashed lines* represent times when the influent NO₃-N concentration and/or the flow rate was changed

3.3.2.2 Nitrite Variations

During operation over the 45 PVs, the influent NO_2-N concentration (mean $\pm$ SD) was 0.02 ± 0.01 mg/L for Column 1, and 0.01 ± 0.01 mg/L for Column 2; whereas the concentrations of NO_2-N in the effluents for Columns 1 and 2 were 0.04 ± 0.02 and 0.02 ± 0.01 mg/L throughout all phases, respectively (Fig. 3.4a, b; Phases I, II, III, IV, V, VI, VII, VIII, and IX). Therefore the differences in influent and effluent nitrite were not significant. The effluent NO_2-N concentration reached a

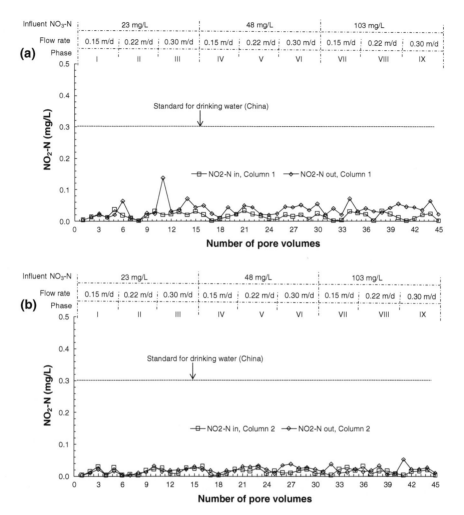

Fig. 3.4 Variations in NO$_2$-N over time under variable operating conditions in **a** Column 1 and **b** Column 2. *Vertical dashed lines* represent times when the influent NO$_3$-N concentration and/or the flow rate was changed

maximum of 0.14 mg/L at PV 11 for Column 1, and 0.05 mg/L at PV 41 for Column 2 (Fig. 3.4a, b). Obviously, both HAD PRB columns are capable of maintaining the NO$_2$-N concentrations in their effluents below the Chinese drinking water standard of 0.3 mg/L. NO$_2$-N did not accumulate in any of the 9 phases of operation in both columns, indicating that there was no competition between nitrite and nitrate reductase for electron donors, and the reduction rate of nitrate was higher than that of nitrite. Nitrite is an intermediate product of nitrate reduction and its accumulation is often observed in pilot-scale studies in BD. NO$_2$-N accumulation is thought to be caused by a limited supply of (in)organic carbon and/or non-optimal environmental

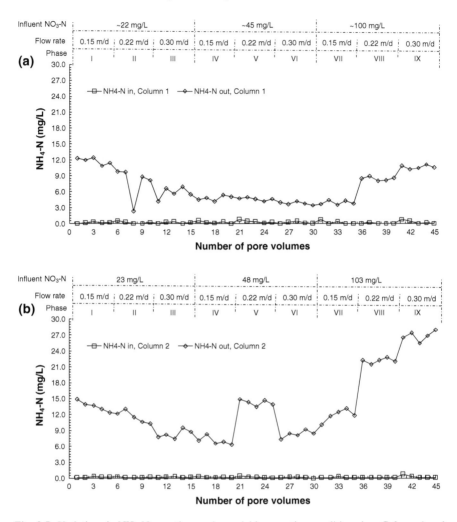

Fig. 3.5 Variations in NH₄-N over time under variable operating conditions in **a** Column 1 and **b** Column 2. *Vertical dashed lines* represent times when the influent NO₃-N concentration and/or the flow rate was changed

conditions. Saliling et al. (2007), for example, reported effluent NO₂-N concentrations around 2.0 mg/L for all bioreactors at a constant flow rate of 15 ml/min and three influent NO₃-N concentrations of 50, 120, and 200 mg/L, which were packed with wood chips, wheat straw and Kaldnes plastic media (made of high-density polyethylene) respectively. Robinson-Lora and Brennan (2009) noted that NO₂-N accumulation tended to reach its maximum (between 5 and 8 mg/L) in a column packed with crab-shell chitin (SC-20) with the onset of incomplete nitrate reduction.

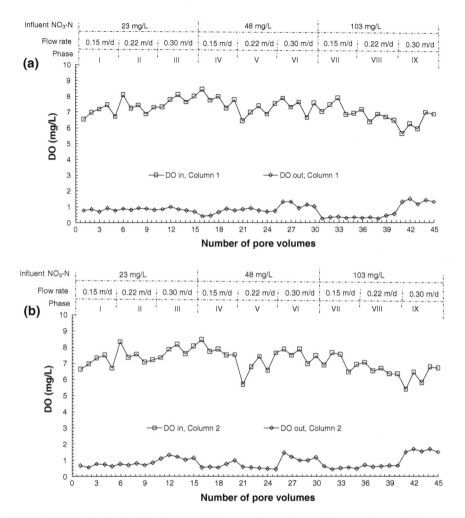

Fig. 3.6 Variations in dissolved oxygen (DO) over time under variable operating conditions in a Column 1 and b Column 2. *Vertical dashed lines* represent times when the influent NO₃-N concentration and/or the flow rate was changed

3.3.2.3 Ammonium Variations

During operation over the 45 PVs, the Column 1 NH₄-N concentration (mean ± SD) in the influent was 0.20 ± 0.28 mg/L, and the Column 2 corresponding value was 0.11 ± 0.19 mg/L; nevertheless, relatively high concentrations were observed in the effluents from Columns 1 and 2 (Fig. 3.5a, b; Phases I, II, III, IV, V,VI, VII, VIII, and IX). A maximum effluent NH₄-N concentration of 12.41 mg/L was observed at PV 3 for Column 1, and up to 28.00 mg/L of effluent NH₄-N was detected at PV 45 for Column 2 (Fig. 3.5a, b). These results indicated

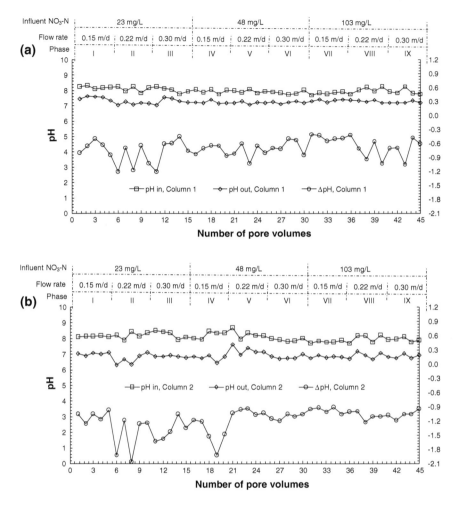

Fig. 3.7 Variations in pH over time under variable operating conditions in **a** Column 1 and **b** Column 2. *Vertical dashed lines* represent times when the influent NO_3-N concentration and/or the flow rate was changed. $\triangle pH = pH_{out} - pH_{in}$

large amounts of NH_4-N were gradually generated. The pine bark, purchased from a local nursery store in Beijing, China, was normally utilized as culture substrate of flowers, grasses and trees. Some ammonium fertilizer had been applied to the pine bark before sale, i.e., the pine bark had been contaminated by the fertilizer. Subsequently, NH_4-N was released to water during this study, causing the high effluent NH_4-N concentration. Additionally, the pine bark used contained organic N, which was deaminated to NH_4-N by ammonifying bacteria under oxygenated conditions (DeSimone and Howes 1998). Another possible source of NH_4-N was that the synthetic chloride form of green rust with an Fe(II): Fe(III) ratio of

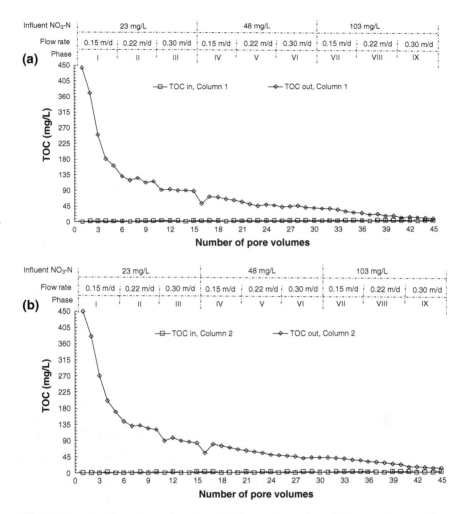

Fig. 3.8 Variations in total organic carbon (TOC) over time under variable operating conditions in **a** Column 1 and **b** Column 2. *Vertical dashed lines* represent times when the influent NO_3-N concentration and/or the flow rate was changed

3: 1 $[Fe_{4.5II}Fe_{1.5III}(OH)_{12}Cl_{1.5}]$ would reduce nitrate to ammonium (Eq. 3.5) (Hansen et al. 2001).

$$Fe_{4.5}^{II}Fe_{1.5}^{III}(OH)_{12}Cl_{1.5} + 5/16NO_3^- \rightarrow 2Fe_3O_4 + 5/16NH_4^+ + 14/16H^+ \\ + 3/2Cl^- + 79/16H_2O \tag{3.5}$$

NH_4-N production would be considered undesirable in both HAD PRB columns. NH_4-N could lead to water quality degradation in the downgradient aquifer. NH_4-N can oxidize back to NO_3-N if aerobic conditions are subsequently encountered farther along the flowpath (Robertson et al. 2008). NH_4-N could also lead to ammonium driven chlorine demand during drinking water treatment, increasing both the cost of treatment and the difficulty in managing chlorine residuals for adequate disinfection.

3.3.2.4 Dissolved Oxygen Variations

During operation over the 45 PVs, the influent DO concentration (mean $\pm$ SD) was 7.19 ± 0.61 mg/L for Column 1, and 7.16 ± 0.69 mg/L for Column 2; however the effluent DO concentration was 0.78 ± 0.33 mg/L for Column 1, and 0.86 ± 0.36 mg/L for Column 2 (Fig. 3.6a, b; Phases I, II, III, IV, V, VI, VII, VIII, and IX). The maximum effluent DO concentrations of 1.48 and 1.69 mg/L were respectively observed at PV 42 for Columns 1 and 2. These demonstrated that most of influent DO was successfully removed by both HAD PRB columns (Fig. 3.6a, b). As mentioned in Chap. 2, granulated spongy iron via chemical reduction (CR) and aerobic heterotrophs via aerobic respiration both contributed to DO removal, and spongy iron played a dominant role in deoxygenation. Combining the results of Figs. 3.3, 3.4, and 3.6, it is evident that DO in the two columns did not have a marked negative effect on NO_3-N removal. Negative effect of oxygen on the heterotrophic and autotrophic denitrifiers was minimized in Columns 1 and 2. Biofilm structure is highly stratified and oxygen has to be transported into biofilms by diffusion (Gómez et al. 2002). A proportion of the oxygen was consumed as an electron acceptor by aerobic heterotrophs in the outer aerobic zone of the biofilm, and furthermore oxygen diffusion throughout the biofilm resulted in a decrease in oxygen concentration corresponding with the biofilm depth (Gómez et al. 2002). Nitrate diffuses through the aerobic zone and then is utilized as another electron acceptor by denitrifying bacteria in an inner anoxic or even anaerobic zone. The dense and thick biofilms around the granulated spongy iron and pine bark media in both columns protected denitrifying bacteria from the oxygen inhibition. Simultaneously, there were sufficient opportunities for heterotrophic denitrifiers to employ nitrate as an electron donor due to the sufficient TOC in Columns 1 and 2 (Fig. 3.8). Therefore, the negative effect of DO was, to some extent, buffered by sufficient TOC. Gómez et al. (2002) similarly concluded that the potential inhibitory effect of DO could be overcome by an incremental increase in the concentration of carbon source.

3.3.2.5 pH Variations

During operation over the 45 PVs, the pH value (mean $\pm$ SD) for the column influent was 8.00 ± 0.17 for Column 1, and 8.09 ± 0.24 for Column 2; the pH value for the column effluent was 7.28 ± 0.14 for Column 1, and 6.90 ± 0.24 for Column 2 (Fig. 3.7a, b; Phases I, II, III, IV, V, VI, VII, VIII, and IX). It was anticipated that the effluent pH would rise compared with the influent pH due to the occurrence of heterotrophic denitrification (HD) (Eq. 3.6), autotrophic denitrification (AD)

(Eq. 3.7), CR of NO_3-N (Eq. 3.8) and DO (Eq. 3.9), anaerobic Fe^0 corrosion (Eq. 3.10) and biological deoxygenation (Eq. 3.11) in these two HAD PRB columns. Unfortunately, the results suggested just the opposite: the effluent pH was 0.40–1.21 units lower than the influent pH for Column 1; the effluent pH was 0.91–2.07 units lower than the influent pH for Column 2 (Fig. 3.7a, b). Soares et al. (2000) also observed pH decrease: the pH of the effluent was approximately 0.5 units lower than that of the influent in a cotton-packed field reactor.

$$C_6H_{10}O_2 + 6NO_3^- + 6H^+ \rightarrow 6CO_2 + 8H_2O + 3N_2 \tag{3.6}$$

$$H_2 + 0.35NO_3^- + 0.35H^+ + 0.052CO_2 \rightarrow 0.17N_2 + 1.1H_2O + 0.010C_5H_7O_2N \tag{3.7}$$

$$5Fe^0 + 2NO_3^- + 6H_2O \rightarrow 5Fe^{2+} + N_2 + 12OH^- \tag{3.8}$$

$$2Fe^0 + O_2 + 2H_2O \rightarrow 2Fe^{2+} + 4OH^- \tag{3.9}$$

$$Fe^0 + 2H_2O \rightarrow H_2 + Fe^{2+} + 2OH^- \tag{3.10}$$

$$2C_6H_{10}O_2 + 15O_2 \rightarrow 12CO_2 + 10H_2O \tag{3.11}$$

$$3Fe^{2+} + Fe^{3+} + Cl^- + 8H_2O \rightarrow Fe_4(OH)_8Cl \text{ (chloride green rust)} + 8H^+ \tag{3.12}$$

$$4Fe^{2+} + 2Fe^{3+} + SO_4^{2-} + 12H_2O \rightarrow Fe_6(OH)_{12}SO_4 \text{(sulfate green rust)} + 12H^+ \tag{3.13}$$

$$4Fe^{2+} + O_2 + 10H_2O \rightarrow 4Fe(OH)_3 + 8H^+ \tag{3.14}$$

In the columns, hydrogen ions formed during the processes of green rust ($Fe_4(OH)_8Cl$ and $Fe_6(OH)_{12}SO_4$) and precipitate ($Fe(OH)_3$) formations (Eqs. 3.12, 3.13, and 3.14) owing to chloride and sulfate ions in the influents (Table 3.2). These processes, together with the bicarbonate in the influent could thus buffer the rise in pH to some extent. Besides, fermentative bacteria may contribute to the decrease in the effluent pH by producing organic acids (Rust et al. 2000).

3.3.2.6 Total Organic Carbon Variations

During operation over the 45 PVs, the influent TOC concentration (mean ± SD) was 1.21 ± 0.64 mg/L for Column 1, and 1.07 ± 0.51 mg/L for Column 2 (Fig. 3.8a, b; Phases I, II, III, IV, V,VI, VII, VIII, and IX). During the same period, as more water passed through the columns decreases in effluent TOC concentration were observed (Fig. 3.8a, b). The TOC concentration in the effluents dropped quickly from 442.43 mg/L at PV 1 to 158.71 mg/L at PV 6 for Column 1, and from 449.88 mg/L at PV 1–169.51 mg/L at PV 6 for Column 2 (Fig. 3.8a, b).

Then it declined gradually to 4.53 mg/L at PV 45 for Column 1, and 10.77 mg/L at PV 45 for Column 2 (Fig. 3.8a, b). The effluent TOC was mainly attributed to the breakdown of complex carbon substrates in pine bark by microorganisms, aided by negligible wash-out of the bacteria in the fixed film reactors (Della Rocca et al. 2005). The surplus TOC suggested that the HD in both HAD PRB columns was not carbon-limited. By combining the results of Figs. 3.3., 3.4, and 3.8, it can be deduced that there seemed to be no inhibition by TOC on nitrate removal. The TOC excess is a common drawback with BD and HAD processes supported by solid organic carbon source. Volokita et al. (1996) found the effluent TOC concentration fluctuated between 0 and 30 mg/L for a cotton HD process. Della Rocca et al. (2006) reported the effluent TOC concentration varied from 6 to 30 mg/L for steel wool and cotton HAD processes. Like NH_4-N production, effluent TOC would also be considered undesirable in the two HAD PRBs because it could also result in water quality degradation in the downgradient aquifer (Robertson et al. 2008). TOC leaching is directly linked to high biomass buildup and further aquifer clogging in that it can allow indigenous microorganisms to reproduce rapidly along the flowpath.

3.3.3 Spatial Variations of Water Quality Indices

The two HAD PRB columns were operated in parallel at the three influent NO_3-N concentrations (23, 46 and 104 mg/L) and a flow rate (0.30 m/d) (corresponding to Phases III, VI, and IX). This was carried out to observe the spatial variations of NO_3-N, NO_2-N, NH_4-N, DO, pH, and TOC along the columns. The experimental results are shown in Figs. 3.9, 3.10, 3.11, 3.12, 3.13, and 3.14.

3.3.3.1 Nitrate Variations

The trends of the NO_3-N reduction upward through the HAD PRB columns were the same for Columns 1 and 2 at the three influent NO_3-N concentrations: most of the influent NO_3-N was removed in the first 25 cm for the low (23 mg/L) and middle (46 mg/L) NO_3-N concentrations, and in the first 65 cm for the high initial concentration (104 mg/L) (Fig. 3.9a, b). Apparently, the denitrifying bacteria existent in the first 25 cm of the two columns did not have the capacity to completely reduce NO_3-N when influent concentration was increased to 104 mg/L. Each column performed better in the lower section (0–65 cm) than in the upper section (65–150 cm). This could be attributed to the fact that the bacterial biomass (particularly those on the microbores of the two media) had fully developed in the lower portions due to sufficient substrates. The column heights (150 cm) were adequately enough to assure the effluent NO_3-N concentration of <3.5 mg/L. The column sizes may be reduced ($h \leq 65$ cm) if 3.5 mg NO_3-N/L is the target final

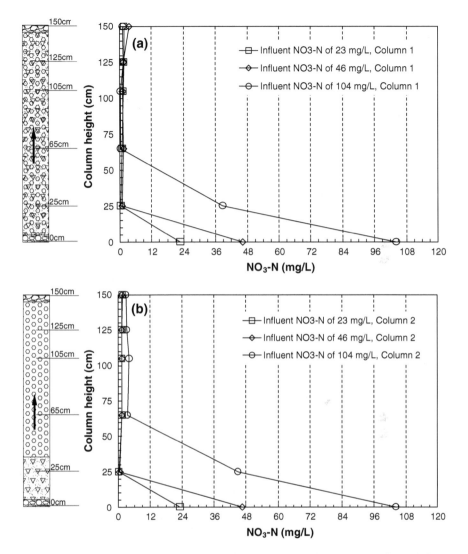

Fig. 3.9 Spatial variations in NO$_3$-N along **a** Column 1 and **b** Column 2 at different influent NO$_3$-N concentrations (at the flow rate of 0.30 m/d). *Arrows* indicate flow direction

effluent concentration. Influent NO$_3$-N concentration and/or flow rate, on the other hand, can be increased if column sizes are not reduced.

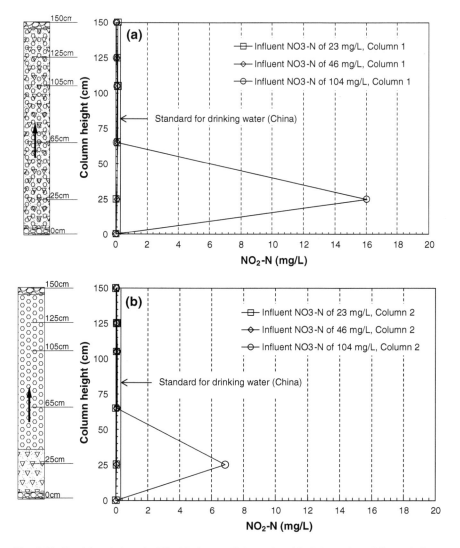

Fig. 3.10 Spatial variations in NO$_2$-N along **a** Column 1 and **b** Column 2 at different influent NO$_3$-N concentrations (at the flow rate of 0.30 m/d). *Arrows* indicate flow direction

3.3.3.2 Nitrite Variations

No NO$_2$-N was generated throughout the columns at the low (23 mg/L) and middle (46 mg/L) influent NO$_3$-N concentrations for Columns 1 and 2 (Fig. 3.10a, b), indicating that the activities of nitrite reductase were greater than (or equal to) those of nitrate reductase, and nitrite reductase was capable of reducing nitrite to gaseous nitrous compounds. However, at the high initial NO$_3$-N concentration

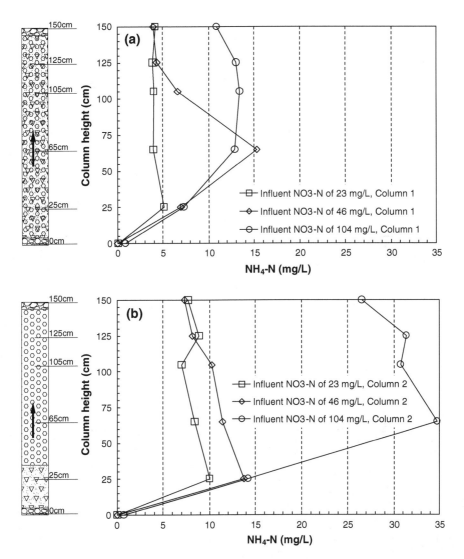

Fig. 3.11 Spatial variations in NH$_4$-N along **a** Column 1 and **b** Column 2 at different influent NO$_3$-N concentrations (at the flow rate of 0.30 m/d). *Arrows* indicate flow direction

(104 mg/L), there was a notable accummulation of NO$_2$-N, 16.04 mg/L for Column 1 and 6.84 mg/L for Column 2, within the first 25 cm of the columns (Fig. 3.10a, b). This showed the introduced NO$_3$-N was quickly transformed to NO$_2$-N in the first 25 cm, but the NO$_2$-N concentration rapidly decreased thereafter to less than 0.1 mg/L at 65 cm for both columns (Fig. 3.10a, b), which was below the Chinese drinking water standard. The data suggested that: (1) the NO$_2$-N reduction mainly occurred between 0 and 65 cm region of the HAD PRB columns;

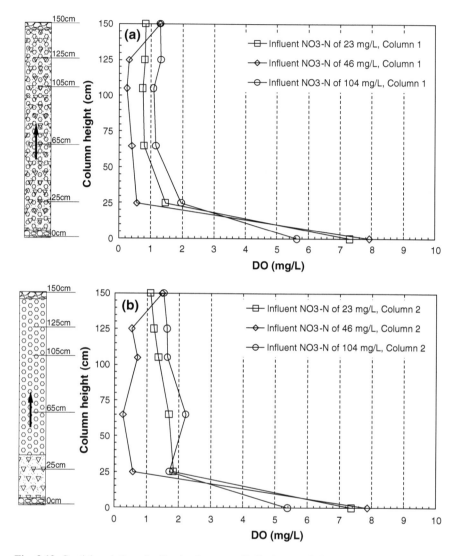

Fig. 3.12 Spatial variations in dissolved oxygen (DO) along **a** Column 1 and **b** Column 2 at different influent NO$_3$-N concentrations (at the flow rate of 0.30 m/d). *Arrows* indicate flow direction

(2) the activities of nitrate reductase were much higher than those of nitrite reductase in the same region; and (3) nitrite reductase was not able reduce all nitrite to gaseous compounds between 0 and 25 cm. Column length of 65 cm was essential to prevent the toxicity of nitrite inhibiting complete water treatment. The NO$_2$-N concentration peak at 25 cm for Column 1 was higher than that for Column 2. This

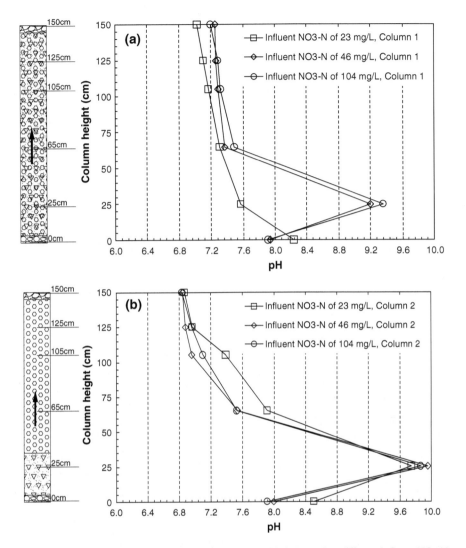

Fig. 3.13 Spatial variations in pH along **a** Column 1 and **b** Column 2 at different influent NO$_3$-N concentrations (at the flow rate of 0.30 m/d). Arrows indicate flow direction

difference was likely due to the lower activities of nitrite reductase in Column 1 Compared with those in Column 2.

3.3.3.3 Ammonium Variations

Contrary to our previous batch studies where no NH$_4$-N was generated (see Chap. 2, Sect. 2.3.2.1), the NH$_4$-N concentrations of 3.5–15.5 mg/L for Column 1 and

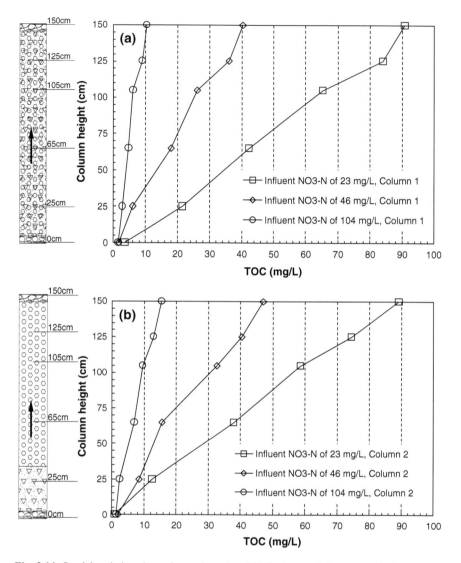

Fig. 3.14 Spatial variations in total organic carbon (TOC) along **a** Column 1 and **b** Column 2 at different influent NO₃-N concentrations (at the flow rate of 0.30 m/d). *Arrows* indicate flow direction

7.0–35.0 mg/L for Column 2 were detected between 25 and 150 cm at the three influent NO₃-N concentrations. The reasons for these phenomena were discussed in Sect. 3.3.2.3. However, the denitrification performance (Fig. 3.9) suggested that the NH₄-N generated (<15.5 mg/L for Column 1 and <35.0 mg/L for Column 2) had no inhibitory effect on hetrotrophic and autotrophic denitrifiers. In contrast, Della Rocca et al. (2006) reported that NH₄-N concentration of 14.62 mg/L inhibited denitrifers in a column packed with steel wool and cotton. Hence, the denitrifiers in the two HAD PRB columns with spongy iron and pine bark performed better than those in the column with steel wool and cotton.

3.3.3.4 Dissolved Oxygen Variations

The DO concentration sharply declined to less than 2 mg/L in the initial 25 cm, and subsequently maintained relatively constant at all the three influent NO_3-N concentrations in Columns 1 and 2 (Fig. 3.12a, b). Clearly, most of the influent DO was removed in the lower sections because of granulated spongy iron and aerobic heterotrophs. By combining the results of Figs. 3.9, 3.10, and 3.12, it was therefore concluded that the influent DO concentration of >5.3 mg/L had little negative effect on the activities of nitrate and nitrite reductase in the two HAD PRB columns. The reasons for this limited effect were discussed in Sect. 3.3.2.4. In contrast, Gómez et al. (2002) observed that the presence of influent DO not only resulted in a decrease in inorganic nitrogen removal, but also in increased nitrite concentration. Combining Figs. 3.9, and 3.12, it can be concluded that NO_3-N reduction and DO removal in the lower sections (0–25 cm) of both columns were parallel processes, which was consistent with a previous study where a continuous, upflow fixed bed reactor was used to remove NO_3-N from drinking water (Mergaert et al. 2001).

3.3.3.5 pH Variations

In a previous study by Della Rocca et al. (2006), pH decreased gradually from initial 8.3 to effluent 6.6 along a column when NO_3-N contaminated water passed through a steel wool zone and then a cotton zone. A similar pH profile along the column height was observed at the influent NO_3-N concentration of 23 mg/L for Column 1 (Fig. 3.13a). At the other two influent concentrations (46 and 104 mg/L) pH peaks appeared at 25 cm and decreased to values between 7.0 and 7.3. The pH increased to a maximum value at 25 cm before gradually decreasing to a value below 7 at 150 cm at any influent NO_3-N concentration in Column 2 (Fig. 3.13b). The rise in pH between 0 and 25 cm could be explained due to the production of hydroxide mainly caused by BD and CR of NO_3-N and DO. The pH at different column heights did not inhibit the performance of both HAD PRB columns because the pH was below 10, which is the tolerance range of bacteria (Su and Puls 2007).

3.3.3.6 Total Organic Carbon Variations

For Column 1, the TOC concentration increased with the column height at the three influent NO_3-N concentrations (Fig. 3.14a). This suggested that the pine bark continuously released TOC, causing TOC accumulation. Within the first 65 cm where NO_3-N removal mainly occurred (Fig. 3.9a), TOC excess was also observed (Fig. 3.14a), indicating that organic carbon did not limit BD, particularly HD. Furthermore, the higher the influent NO_3-N concentration, the lower the TOC concentration (Fig. 3.14a). One explanation was that the more NO_3-N that was removed, the more organic carbon was consumed, the less TOC remained. Another explanation was that the release rates of TOC from the pine bark decreased with operating time increasing. TOC data for Column 2 were comparable to Column 1 (Fig. 3.14a, b).

3.3.4 Effects of Operating Parameters on the Heterotrophic-Autotrophic Denitrification Permeable Reactive Barrier Performance

3.3.4.1 Hydraulic Retention Time

Generally, HRT is an important operating parameter affecting BD efficiency (Moon et al. 2004) because adequate contact time is needed among nitrate, denitrifiers and electron donors. For example, Lee et al. (2010) reported increasing HRT was able to increase hydrogenotrophic denitrificaiton performance in a packed bed reactor. The HRT in Columns 1 and 2 was increased gradually by changing flow rate. The initial HRT was 17.51 d, and then reached values of 11.82 and 8.75 d. The NO_3-N percent removals for Columns 1 and 2 obtained at different HRTs are illustrated in Fig. 3.15.

The high NO_3-N percent removals (97–100 %) for Columns 1 and 2 were achieved at the HRTs ranging from 8.75 to 17.51 d (Fig. 3.15). These results

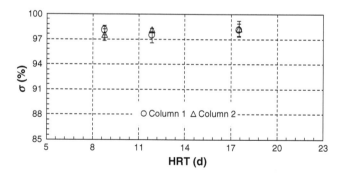

Fig. 3.15 Effect of hydraulic retention time (HRT) on the performance of Columns 1 and 2 at the influent NO_3-N concentration of 104 mg/L. Data points and error bars represent the average of five samples taken under the same operating condition and standard deviation, respectively. NO_3-N percent removal (%): $\sigma = (C_{in} - C_{out})/C_{in} \times 100\%$

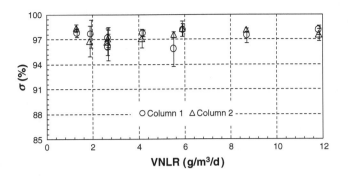

Fig. 3.16 Effect of volumetric NO_3-N loading rate (VNLR) on the performance of Columns 1 and 2. Data points and error bars represent the average of five samples taken under the same operating condition and standard deviation, respectively. NO_3-N percent removal (%): $\sigma = (C_{in} - C_{out})/C_{in} \times 100\,\%$

presented here, suggested that the denitrifiers in the columns were not sensitive to the HRTs and HRT had a negligible effect on the performance of the two HAD PRB columns.

3.3.4.2 Volumetric NO_3-N Loading Rate

VNLR of Columns 1 and 2 was increased gradually by changing influent NO_3-N concentration and flow rate. The initial loading rate was 1.3 $g/m^3/d$, and then reached values of 1.9, 2.7, 4.1, 5.5, 5.9, 8.7 and 11.8 $g/m^3/d$. The NO_3-N percent removals for Columns 1 and 2 obtained at different NO_3-N loading rates are illustrated in Fig. 3.16.

The NO_3-N percent removals for columns 1 and 2 were similar to each other and close to 95–100 %, when VNLRs were in the range of 1.3–11.8 ($g/m^3/d$) (Fig. 3.16). The denitrifiers in the columns had a good ability to adapt to a shock NO_3-N loading, and VNLR had a negligible effect on the performance of the two HAD PRB columns.

3.3.4.3 Packing Structure

Two packing structures were designed in this study (Fig. 3.2). The effect of packing structure on the two HAD PRBs was verified by making comparisons of the effluent NO_3-N, NO_2-N, NH_4-N, DO and TOC concentrations and pH values between Columns 1 and 2. Under different operating conditions (Phases I, II, III, IV, V, VI, VII, VIII, and IX), the concentrations of NO_3-N, NO_2-N, DO and TOC in the effluents for Columns 1 and 2 did not show significant differences (Figs. 3.3, 3.4, 3.6, and 3.8). Even though the high effluent NH_4-N concentration was observed in both columns (Fig. 3.5), NH_4-N concentration (<28.00 mg/L) did not

Table 3.6 Comparison of denitrification capacity among different denitrification processes

Reactive media	System description	Maximum NO$_3$-N percent removal (%)	Maximum volumetric NO$_3$-N loading rate (g/m^3/d)	Water temperature (°C)	Denitrification mechanism	Reference
Ethanol	Reactor study	No[a]	4000	25	HD	Green et al. (1994)
Starch	Reactor study	95–100	460	15	HD	Kim et al. (2002)
Soybean oil	Tank study	39	36	15	HD	Hunter (2001)
Coarse hardwood sawdust (15 %)	Field study	80	2.6	No[a]	HD	Robertson et al. (2000)
Coarse hardwood sawdust + leaf compost + grain seed (15 %)	Field study	83	0.7	No[a]	HD	Robertson et al. (2000)
Coarse hardwood sawdust (20 %)	Field study	91	2.4	No[a]	HD	Robertson et al. (2000)
Coarse wood mulch	Field study	58	1.18–1.77	No[a]	HD	Robertson et al. (2000)
Sawdust (20 %)	PRB	No[a]	0.16–0.29	13–22	HD	Schipper et al. (2004)
Sawdust (30 %)	PRB	No[a]	1.4	12	HD	Schipper et al. (2005)
Cotton	Pilot plant	80–100	360	No[a]	HD	Soares et al. (2000)
Cathodic hydrogen + HCO$_3^-$ +electrolytic Fe powder	Column study	95	0.027[b]	No[a]	AD	Sunger and Bose (2009)
Cathodic hydrogen + HCO$_3^-$ +steel wool + pyrite[c]	Column study	90.0–92.5[b]	1.20–1.23[b]	No[a]	AD	Jha and Bose (2005)
Elemental sulfur + limestone	Column study	~100	~10.19[b]	20	AD	Moon et al. (2008)
Cotton + steel wool (R2)	Column study	>99	0.236	23 ± 5	HD + AD + CR	Della Rocca et al. (2006)
Cotton + steel wool (R3)	Column study	>99	0.258	23 ± 5	HD + AD + CR	Della Rocca et al. (2006)
Cotton + steel wool (R4)	Column study	>99	0.277	23 ± 5	HD + AD + CR	Della Rocca et al. (2006)
Pine bark + spongy iron (Column 1)	Column study	>98.1	11.8	23 ± 5	HD + AD + CR	Current study
Pine bark + spongy iron (Column 2)	Column study	>97.6	11.8	23 ± 5	HD + AD + CR	Current study

[a] No data

[b] Estimated or calculated based on the data given in the article

[c] Pyrite for pH control

show any inhibitory effect on hetrotrophic and autotrophic denitrifiers (Fig. 3.3) and not all NH_4-N in the effluents was from nitrate reduction. Even though there were differences in effluent pH between the two columns (7.28 ± 0.14 for Column 1; 6.90 ± 0.24 for Column 2) (Fig. 3.7), the effluent pH for both columns was below 10 (tolerance range of bacteria) (Fig. 3.7) and thus did not inhibit the performance of the two columns (Fig. 3.3). It can be thus concluded that packing structure had a negligible effect on the two HAD PRB columns.

3.3.5 Comparison of Denitrification Capacity Among Different Denitrification Processes

Table 3.6 presents the comparison of maximum NO_3-N percent removal (MNPR) (%) and maximum volumetric NO_3-N loading rate (MVNLR) ($g/m^3/d$) among different denitrification processes. The MVNLRs ($11.8 \ g/m^3/d$) for the two HAD PRBs in this study were significantly lower than those for the HD processes supported by single formate, ethanol, starch, soybean oil or cotton, but much higher than the values for the other HD, AD and HAD processes given in Table 3.6. The differences are attributed to denitrification mechanism and exper-imental condition. Formate, ethanol, starch and soybean oil which require injection devices and electric power in field applications result in high operating and maintenance costs. Cotton can not be applied on a large scale in that it will be easily and strongly compressed at a relatively high flow rate (Soares et al. 2000). The MNPRs (>97 %) for the HAD PRB processes are ideal, which are equal to or greater than all the other processes (Table 3.6).

3.4 Conclusions

From the results obtained here, the following conclusions can be drawn.

1. Both HAD PRB processes were highly feasible and effective in remediation of nitrate contaminated groundwater. NO_3-N percent removals of >92 % were achieved, and most influent NO_3-N was removed in the first 65 cm.
2. During operation over the 45 PVs, the effluent concentrations of NO_3-N, NO_2-N, NH_4-N and DO were lower than (or equal to) 3.42, 0.14, 12.41 and 1.48 mg/L respectively for Column 1; and the corresponding values were lower than (or equal to) 3.22, 0.05, 28.00 and 1.69 mg/L respectively for Column 2. The pH for the effluent was 7.28 ± 0.14 for Column 1, and 6.90 ± 0.24 for Column 2. The effluent TOC concentration dropped from 442.43 mg/L at PV 1 to 4.53 mg/L at PV 45 for Column 1, and from 449.88 mg/L at PV 1 to 10.77 mg/L at PV 45 for Column 2.

3. Most of the influent NO_3-N was removed in the first 25 cm at the low (23 mg/L) and middle (46 mg/L) NO_3-N concentrations, and in the first 65 cm at the high concentration (104 mg/L) by Columns 1 and 2. No NO_2-N was generated throughout the columns at the low and middle concentrations for Columns 1 and 2, but at the high concentration, there was a notable accummulation of 16.04 mg/L for Column 1 and 6.84 mg/L for Column 2 occurred within the first 25 cm. The NH_4-N concentrations of 3.5–15.5 mg/L for Column 1 and 7.0–35.0 mg/L for Column 2 were detected between 25 and 150 cm at the three influent NO_3-N concentrations.. The DO concentration sharply declined to less than 2 mg/L in the initial 25 cm, and subsequently maintained relatively constant at all the three influent NO_3-N concentrations in Columns 1 and 2. The pH decreased gradually along the column height at the low NO_3-N concentration, but its peaks appeared at 25 cm and decreased to values between 7.0 and 7.3 at the middle and high concentrations in Column 1; meanwhile the pH increased to a maximum value at 25 cm before gradually decreasing to a value below 7 at 150 cm at any influent NO_3-N concentration in Column 2. For Columns 1 and 2, the TOC concentration increased with the column height at the three influent NO_3-N concentrations, and the higher the influent NO_3-N concentration, the lower the TOC concentration.
4. There were negligible effects of HRT, VNLR and packing structure on the performance of the two HAD PRB columns. The high NO_3-N percent removals (97–100 %) for Columns 1 and 2 were achieved at the HRTs ranging from 8.75 to 17.51 d. The percent removals for the two columns were close to 95–100 % at the VNLRs in the range of 1.3–11.8 (g/m^3/d).
5. The MVNLRs (11.8 g/m^3/d) for the two HAD PRBs were significantly lower than those for the HD processes supported by single formate, ethanol, starch, soybean oil or cotton, but much higher than the values for the other HD, AD and HAD processes. The MNPRs (>97 %) for the HAD PRBs are ideal, which are equal to or greater than all the other processes.

References

Ahn SC, Oh SY, Cha DK (2008) Enhanced reduction of nitrate by zero-valent iron at elevated temperatures. J Hazard Mater 156:17–22

Alowitz MJ, Scherer MM (2002) Kinetics of nitrate, nitrite, and Cr(VI) reduction by iron metal. Environ Sci Technol 36:299–306

Darbi A, Viraraghavan T, Butler R, Corkal D (2003) Column studies on nitrate removal from potablewater. Water Air Soil Pollut 150:235–254

Della Rocca C, Belgiorno V, Meric S (2005) Cotton-supported heterotrophic denitrification of nitrate-rich drinking water with a sand filtration post-treatment. Water SA 31:229–236

Della Rocca C, Belgiorno V, Meric S (2006) An heterotrophic/autotrophic denitrification (HAD) approach for nitrate removal from drinking water. Process Biochem 41:1022–1028

DeSimone LA, Howes BL (1998) Nitrogen transport and transformations in a shallow aquifer receiving wastewater discharge: A mass balance approach. Water Resour Res 34(2):271–285

Green M, Tarre S, Schnizer M, Bogdan B, Armon R, Shelef G (1994) Groundwater denitrification using an upflow sludge blanket reactor. Water Res 28(3):631–637

Gómez MA, Hontoria E, González-López J (2002) Effect of dissolved oxygen concentration on nitrate removal from groundwater using a denitrifying submerged filter. J Hazard Mater 90:267–278

Hansen HCB, Guldberg S, Erbs M, Koch CB (2001) Kinetics of nitrate reduction by green rusts-effects of interlayer anion and Fe(II): Fe(III) ratio. Appl Clay Sci 18:81–91

Huang CP, Wang HW, Chiu PC (1998) Nitrate reduction by metallic iron. Water Res 32:2257–2264

Hunter WJ (2001) Use of vegetable oil in a pilot-scale denitrifying barrier. J Contam Hydrol 53:119–131

Jha D, Bose P (2005) Use of pyrite for pH control during hydrogenotrophic denitrification using metallic iron as the ultimate electron donor. Chemosphere 61:1020–1031

Kim YS, Nakano K, Lee TJ, Kanchanatawee S, Matsumura M (2002) On-site nitrate removal of groundwater by an immobilized psychrophilic denitrifier using soluble starch as a carbon source. J Biosci Bioeng 93:303–308

Lee JW, Lee KH, Park KY, Maeng SK (2010) Hydrogenotrophic denitrification in a packed bed reactor: effects of hydrogen-to-water flow rate ratio. Bioresour Technol 101:3940–3946

Mergaert J, Boley A, Cnockaert MC, Müller W-R, Swings J (2001) Identity and potential functions of heterotrophic bacterial isolates from a continuous-upflow fixed-bed reactor for denitrification of drinking water with bacterial polyester as source of carbon and electron donor. System Appl Microbiol 24:303–310

Moon HS, Ahn K-H, Lee S, Nam K, Kim JY (2004) Use of autotrophic sulfur-oxidizers to remove nitrate from bank filtrate in a permeable reactive barrier system. Environ Pollut 129:499–507

Moon HS, Shin DY, Nam K, Kim JY (2008) A long-term performance test on an autotrophic denitrification column for application as a permeable reactive barrier. Chemosphere 73:723–728

Robertson WD, Cherry JA (1995) In situ denitrification of septic-system nitrate using reactive porous media barriers: field trials. Ground Water 33(1):99–111

Robertson WD, Blowes DW, Ptacek CJ, Cherry JA (2000) Long-term performance of in situ reactive barriers for nitrate remediation. Ground Water 38:689–695

Robertson WD, Vogan JL, Lombardo PS (2008) Nitrate removal rates in a 15-year-old permeable reactive barrier treating septic system nitrate. Ground Water Monit Remediat 28:65–72

Robinson-Lora MA, Brennan RA (2009) The use of crab-shell chitin for biological denitrification: batch and column tests. Bioresour Technol 100:534–541

Rust CM, Aelion CM, Flora JRV (2000) Control of pH during denitrification in subsurface sediment microcosms using encapsulated phosphate buffer. Water Res 34:1447–1454

Rust CM, Aelion CM, Flora JRV (2002) Laboratory sand column study of encapsulated buffer release for potential in situ pH control. J Contam Hydrol 54:81–98

Saliling WJB, Westerman PW, Losordo TM (2007) Wood chips and wheat straw as alternative biofilter media for denitrification reactors treating aquaculture and other wastewaters with high nitrate concentrations. Aquacult Eng 37:222–233

Schipper LA, Vojvodić-Vuković M (1998) Nitrate removal from groundwater using a denitrification wall amended with sawdust: field trial. J Environ Qual 27:664–668

Schipper LA, Barkle GF, Hadfield JC, Vojvodić-Vuković M, Burgess CP (2004) Hydraulic constraints on the performance of a groundwater denitrification wall for nitrate removal from shallow groundwater. J Contam Hydrol 69:263–279

Schipper LA, Barkle GF, Vojvodić-Vuković M (2005) Maximum rates of nitrate removal in a denitrification wall. J Environ Qual 34:1270–1276

Soares MIM, Abeliovich A (1998) Wheat straw as substrate for water denitrification. Water Res 32:3790–3794

Soares MIM, Brenner A, Yevzori A, Messalem R, Leroux Y, Abeliovich A (2000) Denitrification of groundwater: pilot-plant testing of cotton-packed bioreactor and post-microfiltration. Wat Sci Technol 42:353–359

Su C, Puls RW (2007) Removal of added nitrate in cotton burr compost, mulch compost, and peat: mechanisms and potential use for groundwater nitrate remediation. Chemosphere 66:91–98

Sunger N, Bose P (2009) Autotrophic denitrification using hydrogen generated from metallic iron corrosion. Bioresour Technol 100:4077–4082

Volokita M, Belkin S, Abeliovich A, Soares MIM (1996) Biological denitrification of drinking water using newspaper. Water Res 30:965–971

Westerhoff P, James J (2003) Nitrate removal in zero-valent iron packed columns. Water Res 37:1818–1830

Yang GCC, Lee HL (2005) Chemical reduction of nitrate by nanosized iron: kinetics and pathways. Water Res 39:884–894

Chapter 4
Bacterial Community in the Inoculum

Abstract Although our previous studies indicated the two heterotrophic-autotrophic denitrification permeable reactive barriers (HAD PRBs) contained heterotrophic and autotrophic denitrifying bacteria and aerobic heterotrophs, convincing molecular and biochemical evidence for their existence is lacking and the bacterial communities remain largely unknown. Using polymerase chain reaction (PCR) and 16S rRNA, the bacterial community composition in the inoculum introduced into the two HAD PRBs were assessed in this study. The extracted deoxyribonucleic acid (DNA) fragment of about 23 kb in length indicated integral genomic DNA was successfully achieved. The A_{260}/A_{280} ratio of approximately 1.72 suggested the genomic DNA could be directly used for subsequent PCR amplification. The 27F/1492R primer pair was successfully able to obtain an approximately 1500-bp specific band. The inoculum contained aerobic heterotrophic bacteria (belonging to *Adhaeribacter* and *Flavisolibacter*), heterotrophic denitrifiers (belonging to *Bacillus*, *Clostridium*, *Flavobacterium*, *Steroidobacter* and *Novosphingobium*), hydrogenotrophic denitrifiers (belonging to *Pseudomonas*) and the other anaerobic bacteria (belonging to *Anaerovorax*, *Azoarcus*, *Geobacter* and *Desulfobulbu*). The diversity of bacteria from the inoculum was high, with at least 13 bacterial genera present.

Keywords: Heterotrophic-autotrophic denitrification · Permeable reactive barriers · Bacterial community · Polymerase chain reaction · 16S rRNA · Aerobic heterotrophic bacteria · Heterotrophic denitrifiers · Hydrogenotrophic denitrifiers

Abbreviations

BD	Biological denitrification
DNA	Deoxyribonucleic acid
dNTP	deoxynucleoside triphosphate
HAD	Heterotrophic-autotrophic denitrification
IPTG	Isopropyl-D-thiogalactopyranoside
LB	Luria–Bertani
MEGA	Molecular evolutionary genetics analysis

F. Liu et al., *Study on Heterotrophic-Autotrophic Denitrification Permeable Reactive Barriers (HAD PRBs) for In Situ Groundwater Remediation*, SpringerBriefs in Water Science and Technology, DOI: 10.1007/978-3-642-38154-6_4, © The Author(s) 2014

NCBI   National Center for Biotechnology Information
PCR   polymerase chain reaction
PRB   Permeable reactive barrier
RDP   Ribosomal Database Project
X-Gal   5-bromo-4-chloro-3-indolyl-D-galactopyranoside

4.1 Introduction

Biological denitrification (BD) is a process in which nitrate is reduced stepwisely to nitrogen gas via nitrite, nitric oxide and nitrous oxide in oxygen-limiting environments (Lim et al. 2005; Dang et al. 2009). Recent studies have shown that BD is mediated by a number of denitrifying bacterial genera, among which are *Pseudomonas, Paracoccus, Flavobacterium, Alcaligenes* and *Bacillus spp*, etc. Most denitrifying bacteria are heterotrophic and are able to utilize a wide range of organic carbon compounds (sugars, organic acids, amino acids, etc.) as sources of electron donors (Hiscock et al. 1991). A typical autotrophic denitrifying bacterium is *Thiobacillus denitrificans* (Hiscock et al. 1991). Communities of denitrifying bacteria in soil (Priemé et al. 2002), marine sediments (Braker et al. 2000) and salt marsh sediments (Cao et al. 2008) have been widely studied. However, those in heterotrophic-autotrophic denitrification (HAD) processes have received little attention. In our previous studies (see Chaps. 2 and 3), the experimental results indicated the truth that the two HAD permeable reactive barriers (PRBs) supported by spongy iron and pine bark contained heterotrophic and autotrophic denitrifiers as well as aerobic heterotrophs, but convincing molecular and biochemical evidence for their existence is still lacking and the bacterial communities remain largely unknown.

Based on polymerase chain reaction (PCR) and 16S rRNA, the objectives of this study were to: (1) explore the bacterial community structure in the inoculum which was introduced into the two HAD PRBs; (2) construct a phylogenetic tree of 16S rRNA gene sequences; (3) confirm the existence of functional bacteria in the HAD process; and (4) provide useful information on bacterial diversity for better understanding nitrate denitrification mechanisms in the HAD PRBs.

4.2 Materials and Methods

4.2.1 Sample Collection for Gene Analysis

A water sample for 16S rRNA genes was collected from the inoculum bottle described in Chap. 3, Sect. 3.2.2. Specifically, the sample was collected aseptically in a sterile bottle and kept on ice. At the laboratory, the sample was homogenized by inverting the bottle at least 3 times, and 200 ml was centrifuged for 10 min (10,000 × g); the pellet was retained for the total deoxyribonucleic acid (DNA) extraction.

4.2.2 Total DNA Extraction and Polymerase Chain Reaction Amplification of 16S rRNA Genes

Total DNA was extracted from 200 ml of the water sample in duplicate using a FastDNA Spin Kit for Soil (MP Biomedicals, Solon, OH, USA), according to the manufracture's protocol. The extracted DNA was stored at −20 °C for subsequent studies and −80 °C for permanent preservation.

Before PCR analysis, the extracts were diluted (either 1: 1 or 1: 10 or 1: 50) with molecular-grade water to minimize the presence of PCR inhibitors.

16S rRNA genes were amplified for clone library construction by universal 16S primers 27F (5′-AGAGTTTGATCATGGC-3′) and 1492R (5′-TACCTTGT-TACGACTT-3′) (Sangon, Shanghai, China). A 25 µL reaction mixture contained 1 µL of purified DNA template, 5 µL of 1X Go *Taq* Flexi buffer, 2.5 µL of MgCl$_2$ solution, 0.5 µL 5 U/µL *Taq* DNA polymerase, 1 µL of 10 µM each primer, and deionized water. At the same time, sterile deionized water was used as negative controls for PCR. Reaction mixtures were cycled in a Veriti PCR apparatus (Applied Biosystem, Carlsbad, CA, USA) with an initial denaturation at 95 °C for 5 min, followed by 35 cycles of denaturation (95 °C for 30 s), annealing (54 °C for 30 s), extension (72 °C for 60 s), and a final extension step carried out at 72 °C for 10 min (Braker et al. 2001). The PCR products were analysed by 1 % (wt/vol) agarose gel electrophoresis to verify that 16S rRNA genes were amplified from the total DNA. DNA marker D2000 was used to determine the molecular size.

4.2.3 Polymerase Chain Reaction Product Purification, Ligation and Transformation

Successfully amplified products were excised and purified with a PCR Purification Mini Kit (Bio Basic Inc., Markham, Canada) as per the manufacture's protocol. Subsequently, the purified PCR products were ligated into pEASY-T1 vector and transformed into *Escherichia coli* Trans-T1 component cells (TransGen Biotech, Beijing, China) with a pEASY-T1 cloning kit (TransGen Biotech, Beijing, China) according to the manufacture's protocol. The transformed cells were plated on Luria-Bertani (LB) plates containing 5-bromo-4-chloro-3-indolyl-D-galactopyranoside (X-Gal, 20 µg/mL) (Sangon, Shanghai, China), isopropyl-D-thiogalactopyranoside (IPTG, 24 µg/mL) (Sangon, Shanghai, PRC) and 100 µg/mL of ampicillin (Sangon, Shanghai, China).

4.2.4 Cloning and Sequencing

Single clonies were picked up from LB plates to serve as DNA templates. Vector primers RV-M (5′-GAGCGGATAACAATTTCACACAGG-3′) and M13-D (5′-AGGGTTTTCCCAGTCACGACG-3′) were utilized as PCR primers in pEASY-T1 vector (TransGen Biotech, Beijing, China) for screening of positive clones. Each PCR reaction mixture (25 µL) contained 1 µM deoxynucleoside triphosphate (dNTP) mix, 1.5 µM of $MgCl_2$ solution, 5 µL of 1X Go *Taq* Flexi buffer, 0.1 µM each primer, 1.5 U *Taq* DNA polymerase, and deionized water. PCR conditions included initial denaturation (94 °C for 3 min), 30 cycles of denaturation (94 °C for 45 s), annealing (58 °C for 60 s) and extension (72 °C for 45 s), followed by a final extension (72 °C for 10 min) at the end of cycling. The PCR products were examined by 1 % (wt/vol) agarose gel electrophoresis. Marker D2000 was used to determine the molecular size.

Positive clones were transformed to 800 µL mixtures (LB medium 100 µg/mL and ampicillin 100 µg/mL each) (Sangon, Shanghai, China) at 37 °C for 12 h. Amplified DNA was sent to Beijing Liuhe Huada Gene Science and Technology Stock Co., Ltd. (Beijing, China) for sequencing.

4.2.5 Sequence Analysis of 16S rRNA

The 16S rRNA sequences were identified via the Ribosomal Database Project II (RDP II) website (http://rdp.cme.msu.edu) (Altschul et al. 1997; Cole et al. 2009), and then assigned to the new phylogenetically consistent higher-order bacterial taxonomy using the RDP Naive Bayesian rRNA classifier (Version 2.2) (95 % confidence interval). Finally, a total of 77 sequences were submitted to the online RDP Seqmatch program (http://rdp.cme.msu.edu/seqmatch/seqmatch_intro.jsp) to identify related 16S rRNA gene sequences in the RDP GenBank database (Cole et al. 2007).

4.2.6 Phylogenetic Analysis

All the 16S rRNA sequences were aligned using ClustalX program (Version 1.8) (Thompson et al. 1997) available at European Bioinformatics Institute Molecular Biology Laboratory (Hinxton, Cambridge, UK) web server and grouped using DOTUR progam at 98 % sequence similarity cut-off level. Phylogenetic trees were constructed by the neighbor-joining method (Saitou and Nei 1987) (with 1,000 bootstrap replicates) using molecular evolutionary genetics analysis (MEGA) software (Version 4.0) (Tamura et al. 2007).

4.2.7 Nucleotide Sequence Accession Numbers

The 16S rRNA nucleotide sequences determined in this study have been deposited in the National Center for Biotechnology Information (NCBI) GenBank database under the accession numbers JF523546-JF523604.

4.3 Results and Discussion

4.3.1 Total DNA Extraction

Total DNA extraction is a key step of PCR-16S rRNA, because DNA purity and integrality significantly effect the following PCR and diversity analysis. The total DNA extracted from the inoculum is shown in Fig. 4.1.

As expected, the extracted DNA fragment was about 23 kb in length, indicating that integral genomic DNA was successfully achieved. In addition, an A_{260}/A_{280} ratio serves as an indicator of DNA purity (Parzer and Mannhalte 1991). The A_{260}/A_{280} ratio between 1.7 and 2.0 is generally accepted as representative of a high-quality DNA. The A_{260}/A_{280} ratio of approximately 1.72 (data not shown) in this study suggested that the genomic DNA could be directly used for subsequent PCR amplification.

Fig. 4.1 Total DNA extraction: Lane M, DNA marker D2000; Lane 1, total DNA

4.3.2 Polymerase Chain Reaction Amplification of 16S rRNA Genes

The DNA extract was diluted 1: 1, 1: 10 and 1: 50, respectively. The 16S rRNA genes were amplified by the universal 16S primers 27F and 1492R. The PCR-amplified gene fragments in this study were electrophoresed on 1 % (wt/vol) agarose gels (Fig. 4.2).

From Fig. 4.2, the 27F/1492R primer pair was successfully able to obtain an approximately 1500-bp specific band, which was satisfactory and could be used for the next processes.

4.3.3 Bacterial Community Structure and Phylogenetic Analysis Based on a 16S rRNA Gene Library

The community structure of bacteria revealed via PCR-16S rRNA is summarized in percentage of different groups in Fig. 4.3. The 16S rRNA sequences obtained from the inoculum were compared to those available in the RDP GenBank (Table 4.1). A phylogenetic tree of the 16S rRNA gene sequences from the inoculum was constructed (Fig. 4.4). Bootstrap analysis (1,000 replicates) was done to confirm the reliability of the phylogenetic tree.

Fig. 4.2 Electrophoresis analysis of polymerase chain reaction (PCR)-amplified 16S rRNA gene fragments: Lane M, DNA marker D2000; Lane 1, 1 × dilution; Lane 2, 10 × dilution; Lane 3, 50 × dilution; Lane N, negative control

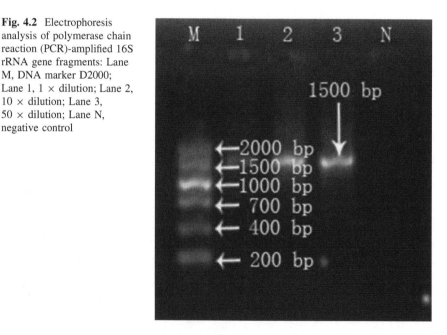

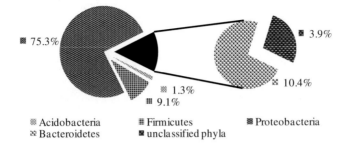

■ 75.3%   ■ 3.9%

                                                               ✕ 10.4%
                          ▒ 1.3%
                          ⧜ 9.1%

▒ Acidobacteria   ⧜ Firmicutes   ▒ Proteobacteria
⋈ Bacteroidetes   ■ unclassified phyla

Fig. 4.3 Bacterial community structure in the inoculum in percentage of 5 different groups

Table 4.1 Comparison of 16S rRNA sequences on a basis of sequence matches

No.	Phylogenetic affiliation	Accession number	Closest match available in the RDP GenBank	Similarity (%)
Acidobacteria (1 sequence)				
1	*Gp6*	JF523590	AY921849	84.4
Firmicutes (7 sequences)				
2	*Anaerovorax*	JF523588	AJ229189	78.9
3	*Clostridium*	JF523558	AJ229224	82.9
4	*Bacillus*	JF523549	HQ433471.1	99
5	Unclassified *genus*	–	–	–
Proteobacteria (58 sequences)				
6	*Desulfobulbus*	JF523567	AJ012591	89.3
7	*Geobacter*	JF523554	EU244079	94.2
8	*Geobacter*	JF523604	U96917	84.4
9	*Azoarcus*	JF523577	Y14701	90.0
10	*Steroidobacter*	JF523557	AY921990	85.2
11	*Pseudomonas*	JF523548	AY247063	98.5
12	*Pseudomonas*	JF523569	X98607	73.1
13	*Pseudomonas*	JF523575	X96788	98.8
14	*Pseudomonas*	JF523599	AF134704	94.4
15	*Novosphingobium*	JF523547	AB025012	88.5
16	*Novosphingobium*	JF523559	AF235994	82.2
17	*Novosphingobium*	JF523564	AB025014	86.8
18	Unclassified *genus*	–	–	–
Bacteroidetes (8 sequences)				
19	*Flavobacterium*	JF523586	AY212593	87.8
20	*Adhaeribacter*	JF523584	EF647593	84.6
21	*Flavisolibacter*	JF523546	AJ863256	90.6
22	Unclassified *genus*	–	–	–
Unclassified *phyla* (3 sequences)				

As shown in Fig. 4.3 and Table 4.1, all 77 sequences were divided into 5 phyla: *Acidobacteria* (1 sequence), *Firmicutes* (7 sequences), *Proteobacteria* (58 sequences), *Bacteroidetes* (8 sequences) and Unclassified *phyla* (3 sequences). At least 13

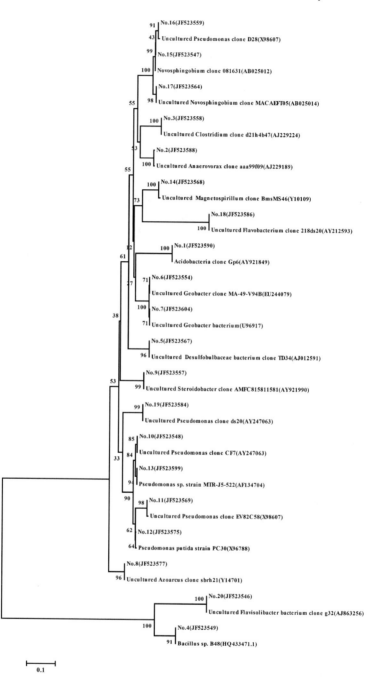

Fig. 4.4 Phylogenetic tree of 16S rRNA gene sequences derived from the inoculum with reference sequences in GenBank. The tree was constructed by the neighbor-joining method using molecular evolutionary genetics analysis (MEGA) software. Bootstrap values (1,000 replicates) are showed at the nodes. The scale bar represents 0.1 substitutions per necleotide position

bacterial genera were identified (Table 4.1), suggesting that bacterial diversity was high. *Proteobacteria* (75.3 %) and *Bacteroidetes* (10.4 %) dominated the bacterial community. The members belonging to *Acidobacteria*, *Firmicutes* and Unclassified *phyla* accounted for 1.3 %, 9.1 % and 3.9 % of the community, respectively.

GP6 is possibly a genus of acidophilic bacteria, which is not well described so far in literature. Naether et al. (2012) reported *GP6* was dominant in grassland soils. *Anaerovorax* is strictly anaerobic genus of fermentative metabolism (Matthies et al. 2000), and has a good ability to ferment putrescine to acetate, butyrate, molecular hydrogen and ammonia (Matthies et al. 2000). *Azoarcus* is a facultatively anaerobic genus including species which fix nitrogen and which anaerobically degrade toluene and other mono-aromatic hydrocarbons (http://www.medicaldictionaryweb.com/Azoarcus-definition). *Geobacter* is an anaerobic genus with the ability to oxidize organic compounds (such as aromatic hydrocarbons) into carbon dioxide while using iron oxides or other metals as electron acceptors (http://www.medicaldictionaryweb.com/Geobacter-definition; http://en.wikipedia.org/wiki/Geobacter). *Desulfobulbu* is a genus of anaerobic sulfate-reducing bacteria which are able to reduce sulfate and sulfide to sulfureted hydrogen. *Adhaeribacter* and *Flavisolibacter* are the genera of aerobic heterotrophs (Rickard et al. 2005; Yoon and Im 2007), which can break down organic carbon into carbon dioxide. *Bacillus*, *Clostridium*, *Flavobacterium*, *Novosphingobium*, *Steroidobacter* and *Pseudomonas* are six genera of containing denitrifying bacteria (Hiscock et al. 1991; Matějů et al. 1992; Lim et al. 2005; Shinoda et al. 2005; Fahrbach et al. 2008; Yuan et al. 2009). Bacteria of the first five genera are considered as heterotrophic denitrifiers (Molongoski and Klug 1976; Gamble et al. 1977; Shinoda et al. 2005; Ayyasamy et al. 2007; Fahrbach et al. 2008; Yuan et al. 2009) because they need organic carbon as energy source and electron donors when reducing nitrate-nitrogen (NO_3–N). *Clostridium* can break down large molecular weight organics into smaller organic acids. Some members of the genus *Flavobacterium* produce almost undetectable quantities of nitrite-nitrogen (NO_2–N) during NO_3–N reduction (Betlach and Tiedje 1981), but they can not use a wide variety of organic carbons (Gamble et al. 1977). *Steroidobacter* that utilizes steroid as carbon source is capable of reducing NO_3–N to N_2O_4 (Fahrbach et al. 2008). *Novosphingobium* is able to transform NO_3–N to NO_2–N, but incapable of transforming NO_2–N to nitrogen gas (Shinoda et al. 2005; Yuan et al. 2009). *Pseudomonas* represents the most common, efficient and active denitrifiers in natural environments (Hiscock et al. 1991; Ayyasamy et al. 2007). *Pseudomonas* as well as *Bacillus* has been found within wastewater, groundwater and soil (Lim et al. 2005; Ayyasamy et al. 2007). *Pseudomonas* is known to contain both hetrotrophic and hydrogenotrophic denitrifying species (Ayyasamy et al. 2007; Gamble et al. 1977; Sahu et al. 2009). However, it can be deduced that the genus *Pseudomonas* contained autotrophic denitrifiers in the inoculum because of the occurrence of autotrophic denitrification (See Sect. 2.3.2.4).

4.4 Conclusion

The diversity of bacteria from the inoculum introduced into the two HAD PRBs was high, with at least 13 bacterial genera present. The inoculum contained aerobic heterotrophic bacteria (belonging to *Adhaeribacter* and *Flavisolibacter*), heterotrophic denitrifiers (belonging to *Bacillus, Clostridium, Flavobacterium, Steroidobacter* and *Novosphingobium*), hydrogenotrophic denitrifiers (belonging to *Pseudomonas*) and the other anaerobic bacteria (belonging to *Anaerovorax, Azoarcus, Geobacter* and *Desulfobulbu*).

References

Altschul SF, Madden TL, Schaffer AA, Zhang J, Zhang Z, Miller W, Lipman DJ (1997) Gapped BLAST and PSI-BLAST: a new generation of protein database search programs. Nucl Acids Res 25(17):3389–3402

Ayyasamy PM, Shanthi K, Lakshmanaperumalsamy P, Lee S-J, Choi N-C, Kim D-J (2007) Two-stage removal of nitrate from groundwater using biological and chemical treatments. J Biosci Bioeng 104(2):129–134

Betlach MR, Tiedje JM (1981) Kinetic explanation for accumulation of nitrite, nitric-oxide, and nitrous-oxide during bacterial denitrification. Appl Environ Microbiol 42(6):1074–1084

Braker G, Ayala-del-Río HL, Devol AH, Fesefeldt A, Tiedje JM (2001) Community structure of denitrifiers, bacteria, and archaea along redox gradients in Pacific Northwest marine sediments by terminal restriction fragment length polymorphism analysis of amplified nitrite reductase (*nirS*) and 16S rRNA genes. Appl Environ Microbiol 67(4):1893–1901

Braker G, Zhou JZ, Wu LY, Tiedje JM (2000) Nitrite reductase genes (*nirK* and *nirS*) as functional markers to investigate diversity of denitrifying bacteria in Pacific Northwest marine sediment communities. Appl Environ Microbiol 66:2096–2104

Cao Y, Green PG, Holden PA (2008) Microbial community composition and denitrifying enzyme activities in salt marsh sediments. Appl Environ Microbiol 74(24):7585–7595

Cole JR, Chai B, Farris RJ, Wang Q, Kulam-Syed-Mohideen AS, McGarrell DM, Bandela AM, Cardenas E, Garrity GM, Tiedje JM (2007) The ribosomal database project (RDP-II): Introducing myRDP space and quality controlled public data. Nucl Acids Res 35:D169–D172

Cole JR, Wang Q, Cardenas E, Fish J, Chai B, Farris RJ, Kulam-Syed-Mohideen AS, McGarrell DM, Marsh T, Garrity GM, Tiedje JM (2009) The ribosomal database project: improved alignments and new tools for rRNA analysis. Nucl Acids Res 37:D141–D145

Dang H, Wang C, Li J, Li T, Tian F, Jin W, Ding Y, Zhang Z (2009) Diversity and distribution of sediment *nirS*-encoding bacterial assemblages in response to environmental gradients in the eutrophied Jiaozhou Bay. China Microb Ecol 58(1):161–169

Fahrbach M, Kuever J, Remesch M, Huber BE, Kämpfer P, Dott W, Hollender J (2008) *Steroidobacter denitrificans* gen. nov., sp. nov., a steroidal hormone-degrading gammaproteobacterium. Int J Syst Evol Microbiol 58:2215–2223

Gamble TN, Betlach MR, Tiedje JM (1977) Numerically dominant denitrifying bacteria from world soils. Appl Environ Microbiol 33:926–939

Hiscock KM, Lloyd JW, Lerner DN (1991) Review of natural and artificial denitrification of groundwater. Water Res 25:1099–1111

Lim Y-W, Lee S-A, Kim SB, Yong H-Y, Yeon S-H, Park Y-K, Jeong D-W, Park J-S (2005) Diversity of denitrifying bacteria isolated from daejeon sewage treatment plant. J Microbiol 43(5):383–390

Matthies C, Evers S, Ludwig W, Schink B (2000) *Anaerovorax odorimutans* gen. nov., sp. nov., a putrescine-fermenting, strictly anaerobic bacterium. Int J Syst Evol Microbiol 50:1591–1594

Matějů V, Čižinská S, Krejčí J, Janoch T (1992) Biological water denitrification: a review. Enzyme Microb Technol 14:170–183

Molongoski OJ, Michael JK (1976) Characterization of anaerobic heterotrophic bacteria isolated from freshwater lake sediments. Appl Environ Microbiol 31(1):83–90

Naether A, Foesel BU, Naegele V, Wüst PK, Weinert J, Bonkowski M, Alt F, Oelmann Y, Polle A, Lohaus G, Gockel S, Hemp A, Kalko EKV, Linsenmair KE, Pfeiffer S, Renner S, Schöning I, Weisser WW, Wells K, Fischer M, Overmann J, Friedrich MW (2012) Environmental factors affect acidobacterial communities below the subgroup level in grassland and forest soils. Appl Environ Microbiol 78(20):7398–7406

Parzer S, Mannhalte C (1991) A rapid method for the isolation of genomic DNA from citrated whole blood. Biochem J 273:229–231

Priemé A, Braker G, Tiedje JM (2002) Diversity of nitrite reductase (*nirK* and *nirS*) gene fragments in forested upland and wetland soils. Appl Environ Microbiol 68:1893–1900

Rickard AH, Stead AT, O'May GA, Lindsay S, Banner M, Handley PS, Gilbert P (2005) *Adhaeribacter aquaticus* gen. nov., sp. nov., a gram-negative isolate from a potable water biofilm. Int J Syst Evol Microbiol 55:821–829

Sahu AK, Conneely T, Nüsslein K, Ergas SJ (2009) Hydrogenotrophic denitrification and perchlorate reduction in ion exchange brines using membrane biofilm reactors. Biotechnol Bioeng 104(3):483–491

Saitou N, Nei M (1987) The neighbor-joining method: A new method for reconstructing phylogenetic trees. Mol Biol Evol 4(4):406–425

Shinoda Y, Akagi J, Uchihashi Y, Lindsay S, Banner M, Handley PS, Gilbert P (2005) Anaerobic degradation of aromatic compounds by *magnetospirillum* strains: isolation and degradation genes. Biosci Biotechnol Biochem 69(8):1483–1491

Tamura K, Dudley J, Nei M, Kumar S (2007) MEGA4: Molecular evolutionary genetics analysis (MEGA) software version 4.0. Mol Biol Evol 24(8):1596–1599

Thompson JD, Gibson TJ, Plewniak F, Plewniak F, Jeanmougin F, Higgins DG (1997) The Clustal_X windows interface: flexible strategies for multiple sequence alignment aided by quality analysis tools. Nucl Acids Res 25(24):4876–4882

Yoon MH, Im WT (2007) *Flavisolibacter ginsengiterrae* gen. nov., sp. nov. and *Flavisolibacter ginsengisoli* sp. nov., isolated from ginseng cultivating soil. Int J Syst Evol Microbiol 57:1834–1839

Yuan J, Lai Q, Zheng T, Shao Z (2009) *Novosphingobium indicum* sp. nov., a polycyclic aromatic hydrocarbon-degrading bacterium isolated from a deep-sea environment. Int J Syst Evol Microbiol 59:2084–2088

Ribosomal Database Project II (1998) Michigan State University, East Lansing. http://rdp.cme.msu.edu. Accessed 3 April 2013

Ribosomal Database Project Seqmatch program (1998) Michigan State University, East Lansing. http://rdp.cme.msu.edu/seqmatch/seqmatch_intro.jsp. Accessed 3 April 2013

Medical Dictionary (2012) U.S. National Library of Medicine, Maryland. http://www.medicaldictionaryweb.com/Azoarcus-definition. Accessed 3 April 2013

Medical Dictionary (2012) U.S. National Library of Medicine, Maryland. http://www.medicaldictionaryweb.com/Geobacter-definition. Accessed 3 April 2013

Wikipedia (2001) Wikimedia Foundation, Los Angeles. http://en.wikipedia.org/wiki/Geobacter. Accessed 3 April 2013

Index

F. Liu et al., *Study on Heterotrophic-Autotrophic Denitrification Permeable Reactive Barriers (HAD PRBs) for In Situ Groundwater Remediation*, SpringerBriefs in Water Science and Technology, DOI: 10.1007/978-3-642-38154-6, © The Author(s) 2014